T0353496

Thermodynamic Models for Chemical Processes

Series Editor
Laurent Falk

Thermodynamic Models for Chemical Processes

Design, Develop, Analyze and Optimize

Jean-Noël Jaubert
Romain Privat

ELSEVIER

First published 2020 in Great Britain and the United States by ISTE Press Ltd and Elsevier Ltd

ISTE Press Ltd
27-37 St George's Road
London SW19 4EU
UK

www.iste.co.uk

Elsevier Ltd
The Boulevard, Langford Lane
Kidlington, Oxford, OX5 1GB
UK

www.elsevier.com

Notices

Knowledge and best practice in this field are constantly changing. As new research and experience broaden our understanding, changes in research methods, professional practices, or medical treatment may become necessary.

Practitioners and researchers must always rely on their own experience and knowledge in evaluating and using any information, methods, compounds, or experiments described herein. In using such information or methods they should be mindful of their own safety and the safety of others, including parties for whom they have a professional responsibility.

To the fullest extent of the law, neither the Publisher nor the authors, contributors, or editors, assume any liability for any injury and/or damage to persons or property as a matter of products liability, negligence or otherwise, or from any use or operation of any methods, products, instructions, or ideas contained in the material herein.

For information on all our publications visit our website at http://store.elsevier.com/

British Library Cataloguing-in-Publication Data
A CIP record for this book is available from the British Library
Library of Congress Cataloging in Publication Data
A catalog record for this book is available from the Library of Congress
ISBN 978-1-78548-209-0

Printed and bound in the UK and US

Contents

Preface

This book is not a course book but a vademecum aimed primarily at users of process simulation software. It provides, in a condensed form, the methods governing the choice of a thermodynamic model for a targeted application and reminds the reader of the essential concepts on which these methods are based. At the time of publication (2020), this book was up to date with the main models of practical interest for the engineer and the researcher.

Finally, because we are convinced that, in order to use these models properly, it is important to understand – at the very least – how to carry out the calculations of the properties of interest that result from them, we will also refer to the major points.

Happy reading!

Jean-Noël JAUBERT
Romain PRIVAT
May 2020

Correlations for the Estimation of Thermodynamic Properties of Pure Substances in the Liquid, Perfect Gas or Vapor–Liquid States

1.1. Introduction

A correlation is a mathematical expression, usually simple, of a physico-chemical property y as a function of one or several variables x. In this chapter, dedicated to pure substances, two types of properties are considered:

– **properties depending on the temperature**: for example, the heat capacity at constant pressure of a liquid pure substance (assumed to be incompressible), the saturated vapor pressure of a pure substance, the density of a liquid pure substance (assumed to be incompressible), etc. In these cases, the variable of the correlations is $x = T$;

– **constant properties (which do not depend on the temperature)**: for example, the normal boiling point, the critical pressure, the acentric factor, etc.

We will look at which principal correlations can be used for the pure-substance thermodynamic properties that depend on the temperature T. This chapter will end with a brief presentation of the methods that allow us to *predict* the constant properties and those depending on T using only knowledge of the chemical structure of pure substances. The variables x of correlations then become structural information (as well as T in the case of properties that depend on the temperature).

The T-dependent properties that are taken into consideration in this chapter are:

– the characteristic properties of the **vapor–liquid equilibrium of a pure substance**: saturated vapor pressure $P^{sat}(T)$, molar enthalpies of a boiling liquid $h_L^{sat}(T)$ and of saturated vapor $h_G^{sat}(T)$, molar enthalpy of vaporization $\Delta_{vap}H(T) = h_G^{sat}(T) - h_L^{sat}(T)$, molar heat capacity at constant pressure of the boiling liquid $c_{P,L}^{sat}(T)$, densities of the boiling liquid $\rho_L^{sat}(T)$ or of the saturated gas $\rho_G^{sat}(T)$;

– properties of a **pure perfect gas**: molar enthalpy $h^{\bullet}(T)$, molar heat capacity at constant pressure $c_P^{\bullet}(T)$;

– properties of a **subcooled liquid**[1]: molar enthalpy $h_L^{pure}(T)$, molar heat capacity at constant pressure $c_{P,L}^{pure}(T)$, volumetric mass $\rho_L^{pure}(T)$. Here, we assume that pure subcooled liquids are incompressible; in other words, their properties are not affected by the pressure.

NOTE.– Properties of a non-perfect gaseous pure substance cannot be described by the correlations presented in this chapter. Suitable methods (mainly based on equations of state) are presented in the following chapters.

1.2. Thermodynamics of the vapor–liquid equilibrium of a pure substance: what should be remembered

1.2.1. *Phase-intensive variables and global-intensive variables*

The intensive state of a two-phase system can be described by two sets of variables:

– **phase**-intensive variables, which are the intensive variables specific to one of the two phases of the system, for example, the temperature of the liquid phase, the pressure of the gas phase, the density of the liquid, the molar heat capacity of the gas phase at constant pressure, etc.;

---

1 The term "subcooled" refers to a liquid that has been brought to a temperature below its boiling point. Consequently, the pure substance is not boiling (is not at vapor–liquid equilibrium).

– **global**-intensive variables that are not specific to a given phase but which instead characterize the two-phase system as a whole (i.e. the whole system). As an example, we cite the *molar proportion of the gas phase* (defined as the quantity of matter present in the gas phase divided by the global quantity of matter in the system, i.e. the sum of the quantities of matter in the liquid and gas phases), the global density (defined as the global mass of the system, i.e. $m_{liquid} + m_{gas}$, divided by the global volume, i.e. $V_{liquid} + V_{gas}$), etc.

NOTE.– As explained in the following, at the vapor–liquid equilibrium, the two phases have the same temperature $(T_L = T_G)$, the same pressure $(P_L = P_G)$ and each component has the same chemical potential in each phase $(\mu_{i,L} = \mu_{i,G})$. Thus, the temperature, pressure and chemical potential of a component i are both phase-intensive and global-intensive variables.

Since the extensive properties are additive, the global extensive properties of a two-phase system are obtained by adding the extensive properties of the two phases:

$$
\underbrace{Q_{\substack{\text{two-phase} \\ \text{system}}}}_{\substack{\text{Global extensive} \\ \text{property}}} = \underbrace{Q_{\text{liquid phase}}}_{\substack{\text{Extensive property} \\ \text{of the liquid phase}}} + \underbrace{Q_{\text{gas phase}}}_{\substack{\text{Extensive property} \\ \text{of the gas phase}}}
$$

$$
with: Q = \begin{cases} n \text{ (quantity of matter in mol)} \\ V \text{ (volume in m}^3\text{)} \\ H \text{ (enthalpy in J)} \\ C_P \text{ (heat capacity in } J \cdot K^{-1}) \\ ... \end{cases} \qquad [1.1]
$$

Consequently, the relationship between the molar properties of the phases that derive from extensive properties ($q_{\text{liquid phase}} = \dfrac{Q_{\text{liquid phase}}}{n_{\text{liquid phase}}}$, $q_{\text{gas phase}} = \dfrac{Q_{\text{gas phase}}}{n_{\text{gas phase}}}$) and global molar properties ($q_{\substack{\text{two-phase} \\ \text{system}}} = \dfrac{Q_{\substack{\text{two-phase} \\ \text{system}}}}{n_{\substack{\text{two-phase} \\ \text{system}}}}$) is:

$$q_{\substack{\text{two-phase} \\ \text{system}}} = \frac{n_{\text{liquid phase}}}{n_{\substack{\text{two-phase} \\ \text{system}}}} q_{\text{liquid phase}} + \frac{n_{\text{gas phase}}}{n_{\substack{\text{two-phase} \\ \text{system}}}} q_{\text{gas phase}} \qquad [1.2]$$

with $q = \{v, h, c_p \ldots\}$. Equation [1.2] introduces the molar proportions of the phases:

$$\left\{ \begin{array}{l} \text{Molar proportion of gas:} \quad \tau = n_{\text{gas phase}} \big/ n_{\substack{\text{two-phase} \\ \text{system}}} \\[2em] \text{Molar proportion of liquid:} \quad 1 - \tau = n_{\text{liquid phase}} \big/ n_{\substack{\text{two-phase} \\ \text{system}}} \end{array} \right. \qquad [1.3]$$

Figure 1.1 represents a two-phase system that illustrates, on the one hand, the notion of extensive (also known as *total*) and intensive properties that are specific to one of the phases, and, on the other hand, the notion of properties that are characteristic of the two-phase system (known as *global*).

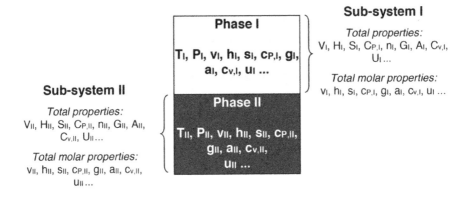

Figure 1.1. *Definition of total, total molar, global and global molar properties*

1.2.2. Conditions of two-phase equilibrium of a pure substance

For **two-phase equilibrium** to be observed, three conditions must be met:

– *thermal equilibrium condition:* there is no heat transfer between the phases in equilibrium, and consequently, these necessarily have the same temperature:

$$T_I = T_{II} \qquad\qquad [1.4]$$

– *mechanical condition of equilibrium:* there is no work transfer of the pressure forces between the phases in equilibrium, and consequently, these are necessarily at the same pressure:

$$P_I = P_{II} \qquad\qquad [1.5]$$

– *diffusive equilibrium condition:* there is no transfer of matter between the phases in equilibrium, and consequently, the chemical potential of the pure substance g (molar Gibbs energy) is the same in both phases:

$$g_I = g_{II}. \qquad\qquad [1.6]$$

1.2.3. Why should the intensive properties of the liquid and vapor phases of a pure substance in vapor–liquid equilibrium be seen as temperature functions?

The answer to this question is provided by the Gibbs phase rule described below. This theorem concerns the variance of a thermodynamic system, the definition of which is also provided.

According to the previous paragraph, the intensive variables of a phase α are the temperature T_α, the pressure P_α and the total molar properties y_α ($y \in \{v, h, g, s ...\}$).

DEFINITION.– The *variance* is defined as the number of independent INTENSIVE variables of PHASES that need to be set (specified) in order to characterize the INTENSIVE variables of ALL PHASES.

The **Gibbs phase rule** is a theorem asserting that the variance of a thermodynamic system that contains one or several phases in equilibrium, with no other characteristics, is given by the expression:

$$v = c + 2 - \varphi \quad \text{with:} \begin{cases} v = \text{variance of the system} \\ c = \text{number of components} \\ \varphi = \text{number of phases in equilibrium} \end{cases} \quad [1.7]$$

Consequently, the variance of a two-phase system ($\varphi = 2$) that contains a pure substance ($c = 1$) is $v = 1$. This means that by specifying a single intensive variable of one out of the two phases, and the value of all the others is then set.

For example, let us suppose that the pressure P_G of the gas phase of a pure substance in vapor–liquid equilibrium is specified. Because of the Gibbs phase rule, this then necessarily sets the pressure of the other phase in equilibrium (this is obvious as the condition of equilibrium between phases imposes $P_L = P_G$), the temperature common to the two phases $T_L = T_G$ (common according to the phase-equilibrium condition), the molar volumes v_L^{sat} and v_G^{sat} of the phases, the molar enthalpies h_L^{sat} and h_G^{sat} of the phases, etc.

A two-phase equilibrium is therefore monovariant. This feature induces an immediate graphical consequence: **in the plane projections of phase diagrams for pure substances** with two intensive phase variables as X and Y coordinates (e.g. pressure–molar volume, pressure–temperature, molar enthalpy–molar entropy planes), two-phase equilibria are represented by curves (the curves are the graphical representations of single-variable functions).

Similarly, the Gibbs phase rule, which predicts that the variance of a pure single-phase substance ($\varphi = 1$) is equal to 2 and that the variance for a pure three-phase substance is zero, the associated graphical representations in the plane projections of the phase diagrams will be surface regions and single points respectively. These remarks are summarized in Figure 1.2 (the definition of the critical point, which is represented in this figure, will be provided at a later stage).

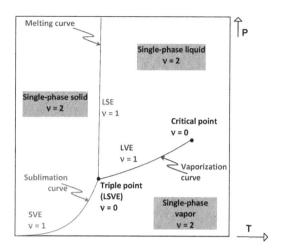

Figure 1.2. *Projection of the phase diagram for a pure substance in the pressure–temperature plane. Illustration of the relationship between the variance of a system and the nature of its representation on the plane. VLE = vapor–liquid equilibrium, LSE = liquid–solid equilibrium, SVE = solid–vapor equilibrium, LSVE = liquid–solid–vapor equilibrium*

Let us now return to the example of a vapor–liquid pure substance for which we specified the pressure of the gas phase. We will illustrate on a graph the monovariant nature of this two-phase equilibrium. Figure 1.3 is a representation in the (P, v) and (P, T) plane projections of the phase diagram of a given pure substance. The specification of the pressure P_G of the gas phase in vapor–liquid equilibrium is shown by a horizontal dot-dash straight line.

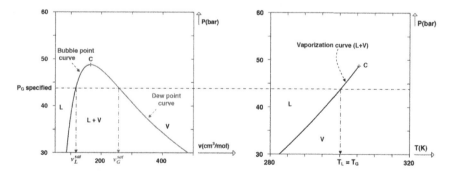

Figure 1.3. *Illustration of the Gibbs phase rule in the case of a pure substance in vapor–liquid equilibrium for which the pressure of the gas phase is specified. Left: phase diagram of a pure substance in the plane (P, v). Right: phase diagram in the plane (P, T). "L": single-phase liquid domain, "V": single-phase vapor domain, "L+V": two-phase vapor–liquid domain; "C": critical point*

The (P,v) plane shows that by specifying P_G, the following properties are then defined: (i) a unique molar volume for the liquid phase (v_L^{sat}) defined by the intersection point of the bubble point curve (representing the vapor–liquid equilibrium pressure versus v_L^{sat}) and the horizontal straight line $P = P_G$, (ii) a unique molar volume for the gas phase (v_G) found at the intersection of the dew point curve (representing the vapor–liquid equilibrium pressure versus v_G^{sat}) and the horizontal straight line $P = P_G$. The (P,T) plane, on the other hand, shows that this specification induces also a unique vapor–liquid equilibrium temperature $T_L = T_G$ (at the intersection of $P = P_G$ and the vaporization curve).

NOTE.– The temperature of a pure substance in vapor–liquid equilibrium is known as the **boiling point temperature**. When the pressure is specified, it is noted $T_{bp}(P)$. The **normal boiling point temperature**, denoted as T_{bp}°, is the boiling point temperature under normal pressure $P^{\circ} = 1\,\text{atm}$, in other words 101,325 Pa.

Choice of the variable for correlations of phase-intensive properties for pure substances in vapor–liquid equilibrium: the previous paragraph insists on the need to set an intensive variable of a phase in order to characterize a vapor–liquid pure substance. In practice, often the temperature is chosen as the correlation variable. In doing so, the intensive properties of the phases are written: $P_L(T) = P_G(T)$ [pressures], $h_G^{sat}(T)$, $h_L^{sat}(T)$ [molar enthalpies], $v_L^{sat}(T)$, $v_G^{sat}(T)$ [molar volumes], $g_L^{sat}(T) = g_G^{sat}(T)$ [molar Gibbs energies], etc.

NOTE.– The pressure of a pure substance in vapor–liquid equilibrium is known as the **saturated vapor pressure**. When the temperature is specified, it is written $P^{sat}(T)$. The molar volumes of the liquid and gas phases in vapor–liquid equilibrium are known as **molar volumes of the saturated liquid** and **saturated gas** phases, denoted respectively as $v_L^{sat}(T)$ and $v_G^{sat}(T)$.

The definitions of the boiling point temperature (at a specified pressure) and of the saturated vapor pressure (at a specified pressure) are summarized in Figure 1.4.

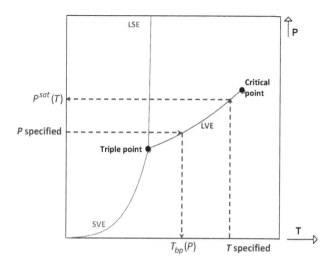

Figure 1.4. *Illustration of the definitions of the saturated vapor pressure and of the boiling point temperature in the pressure–temperature plane of a pure substance. VLE = vapor–liquid equilibrium, LSE = liquid–solid equilibrium, SVE = solid–vapor equilibrium*

1.2.4. *Critical point of a pure substance*

The critical point of a pure substance can be defined in three ways:

– in the (P, T) plane, it is the end point of the vaporization curve of a pure substance (see Figure 1.2);

– in the (P, v) plane, this is the common maximum for the bubble point and dew point curves (see Figure 1.3);

– on the critical isotherm in the (P, v) plane, it is a point of inflection with a horizontal tangent. This definition of the critical point is illustrated in the following section.

1.2.5. *Isotherms of a pure species in the fluid region*

In the (P, v) plane, an isotherm for a pure substance is a curve made up of all the (molar volume, pressure) couples associated with the same temperature.

***Subcritical* isotherms** (subcritical means that the temperature is below the critical point temperature) are made up of three sections:

– **a two-phase vapor–liquid domain:** when the temperature is between the triple and critical temperatures, the pure substance has a unique vapor–liquid equilibrium pressure known as *saturated vapor pressure*. In other words, at a fixed temperature, all the vapor–liquid states of a pure substance are isobaric; consequently, on an isothermal curve, a horizontal plateau (which shows states of the same pressure) characterizes the vapor–liquid equilibrium of the pure substance;

– **a single-phase liquid domain:** liquids under moderate pressure have the feature of being nearly *incompressible* (i.e. their molar volumes do not depend on pressure); as a result, the liquid single-phase domain of a subcritical isotherm can be roughly assimilated to a vertical straight line in the (P, v) plane (liquid states are characterized by a same molar volume);

– **a single-phase vapor domain:** in the case of a pure substance under low pressure (typically < 5 bar), a gas phase can be assimilated to a perfect gas so that the pressure is given by the equation $P = RT/v$. In the (P, v) plane, the single-phase gas region of a subcritical isotherm can be assimilated to a branch of a hyperbolic curve.

By way of an illustration, two subcritical isotherms of ethane at $T = 290$ K and $T = 300$ K are represented in the (P, v) plane in Figure 1.5 (left); on the right, we show the (P, T) plane for the pure substance in which the isotherms are vertical straight lines.

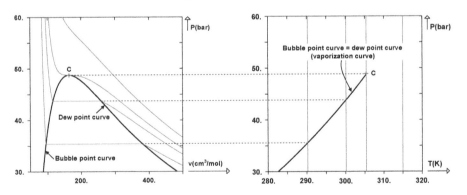

Figure 1.5. *Representation of four isotherms (thin lines) of pure ethane in the (pressure, molar volume) plane (left) and (pressure, temperature) plane (right). C = critical point. The thick lines represent the intensive states of the phases in vapor–liquid equilibrium (bubble point curve, dew point curve, vaporization curve)*

As the temperature is increased, the horizontal vapor–liquid equilibrium plateau in the (P,v) plane reduces: the molar volumes of the liquid and vapor phases in equilibrium move closer together (see Figure 1.5, left).

At the critical point C, these two phases have exactly the same molar volumes. In a more general way, at the critical point, the intensive properties of the liquid phase become identical to those in the gas phase. A *critical isotherm* is an isotherm drawn at the critical temperature of a pure substance (see the isotherm passing through C in Figure 1.5). On this curve (Figure 1.5, left), the vapor–liquid equilibrium plateau present at subcritical temperatures is reduced to a single point C. Geometrically, this induces the existence of an inflection point with a horizontal tangent on C. Mathematically, a critical point is characterized by the following two equations:

$$\begin{cases} (\partial P / \partial v)_T \big|_{\substack{T=T_c \\ v=v_c}} = 0 \\ (\partial^2 P / \partial v^2)_T \big|_{\substack{T=T_c \\ v=v_c}} = 0 \end{cases} \tag{1.8}$$

The quantity v_c is called the *critical molar volume of the pure substance.*

Supercritical **isotherms:** since the critical point has been previously defined as the end point of the vaporization curve in the (P,T) plane, it follows that at all temperatures higher than T_c, the critical temperature, the pure substance remains single phase at all pressures; the isotherm in the (P,v) plane does not have a plateau associated with a change of state and looks approximately like a hyperbolic curve.

1.2.6. *Property changes on vaporization*

For all temperature ranging between the triple and critical temperatures, the molar property changes on vaporization are defined by:

$$\Delta_{vap} X(T) = x_G^{sat}(T) - x_L^{sat}(T) \tag{1.9}$$

where x_G^{sat} and x_L^{sat} denote the molar properties of the gas and liquid phases in vapor–liquid equilibrium respectively.

As an example, the molar volume of vaporization and the molar enthalpy of vaporization[2] are defined by:

$$\begin{cases} \Delta_{vap}V(T) = v_G^{sat}(T) - v_L^{sat}(T) \\ \Delta_{vap}H(T) = h_G^{sat}(T) - h_L^{sat}(T) \end{cases}$$

[1.10]

When the temperature becomes critical, the molar properties of the liquid and gas phases become identical; as a result, the properties of vaporization cancel each other out:

$$\Delta_{vap}X(T_c) = 0$$

[1.11]

1.3. Correlations for the saturated vapor pressure of pure substances

1.3.1. *Practical expressions for the saturated vapor pressure deduced from the Clapeyron equation*

The Clapeyron equation is an exact relationship between the derivative of the saturated vapor pressure with respect to the temperature, on the one hand, and the enthalpy and volume of vaporization, on the other hand:

$$\frac{dP^{sat}}{dT} = \frac{\Delta_{vap}H(T)}{T \times \Delta_{vap}V(T)} = \frac{h_G^{sat}(T) - h_L^{sat}(T)}{T \times \left[v_G^{sat}(T) - v_L^{sat}(T) \right]}$$

[1.12]

The temperature T is the absolute temperature (expressed in kelvin, for example). This equation can be used to deduce a mathematical expression for the $P^{sat}(T)$ function. First, let us recall that the molar compressibility factor is defined as:

$$z = \frac{Pv}{RT}$$

[1.13]

---

2 In practice, the molar adjective is often left out.

Consequently, the (molar) compressibility factor of vaporization is given by the expression:

$$\Delta_{vap}Z(T) = \frac{P^{sat}(T) \cdot \Delta_{vap}V(T)}{RT}$$
[1.14]

The Clapeyron equation [1.12] can be rewritten using the quantity $\Delta_{vap}Z(T)$:

$$\frac{d \ln P^{sat}}{dT} = \frac{\Delta_{vap}H(T)}{RT^2 \times \Delta_{vap}Z(T)}$$
[1.15]

Or equivalently:

$$\frac{d \ln P^{sat}}{d(1/T)} = \underbrace{-\frac{1}{R} \times \frac{\Delta_{vap}H(T)}{\Delta_{vap}Z(T)}}_{\text{denoted } B} = -B$$
[1.16]

This equation gives rise to a quantity B. Experimental measurements show that as a first approximation, B can be considered as constant. This statement is supported by Figure 1.6, which demonstrates the relative constancy of the B function with respect to the variations of the $\Delta_{vap}H(T)$ function.

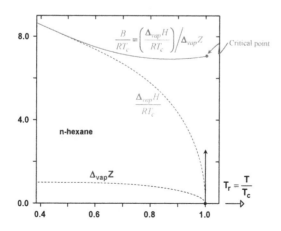

Figure 1.6. *Representation of the temperature variations of the adimensional functions of pure n-hexane $\Delta_{vap}H(T)/(RT_c)$, $\Delta_{vap}Z(T)$ and their ratio, denoted as $B/(RT_c)$*

By integrating the relationship [1.16] with respect to the variable $1/T$ and assuming a constant value of B, we obtain:

$$\ln P^{sat}(T) = A - \frac{B}{(T/\text{K})} \qquad [1.17]$$

where A is a constant produced by the integration and (T/K) is the absolute temperature expressed in kelvin (we point out that another unit of absolute temperature such as the Rankine scale would also have been suitable). Equation [1.17] is called the *integrated form of the Clapeyron equation*. This expression, involving two adjustable parameters (A and B) is indeed suitably adapted in practice to describe the saturated vapor pressures of pure substances over restricted temperature ranges. When this formula is applied to the entire domain $\left[T_{triple} ; T_c \right]$, differences between the experimental and calculated points increase as the size of the molecule increases, as illustrated in Figure 1.7.

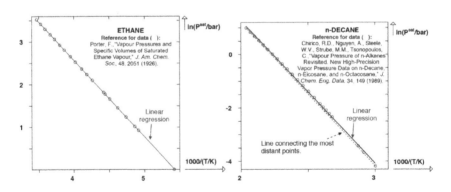

Figure 1.7. *Representation of the natural logarithm of the saturated vapor pressure as a function of the inverse temperature in the case of ethane (left) and of n-decane (right). (O) experimental points; continuous lines: correlation*

To determine the two parameters (A and B) of equation [1.17], it is sufficient to have two couples of points $\left(1/T ; \ln P^{sat} \right)$. Free or commercial experimental databanks contain various types of data that allow determination of the two coefficients A and B. Some illustrations are given as follows:

– Let us suppose that we have the following data: (i) the normal boiling point temperature T°_{bp}, which is the boiling point temperature under a pressure of $P^\circ = 1\,atm$, (ii) critical coordinates (T_c, P_c). If we assume that equation [1.17] is valid, the coefficients A and B are then the solutions to the system:

$$\begin{cases} \ln P^\circ = A - B / T^\circ_{bp} \\ \ln P_c = A - B / T_c \end{cases} \qquad [1.18]$$

After resolution of the system [1.18], we obtain expressions for the coefficients A and B that can be inserted into equation [1.17]. When all calculations have been done, the following correlation is obtained:

$$\ln\left(\frac{P^{sat}}{P^\circ}\right) = \ln\left(\frac{P_c}{P^\circ}\right) \times \left[\frac{(T_c / K)}{(T_c / K) - (T^\circ_{bp} / K)}\right]\left[1 - \frac{(T^\circ_{bp} / K)}{(T / K)}\right] \qquad [1.19]$$

– Let us now suppose that we dispose of the data: (i) (T_c, P_c) and (ii) acentric factor ω. The definition of this property is provided below:

DEFINITION.– The acentric factor of a pure substance is defined by:

$$\omega \underset{\text{déf}}{=} -\log_{10}\left(\frac{P^{sat}(T = 0.7T_c)}{P_c}\right) - 1$$

From the definition of the acentric factor, we deduce that the saturated vapor pressure of a pure substance at the temperature $0.7T_c$ is equal to $P_c \cdot 10^{-(\omega+1)}$. As previously seen, accepting the validity of equation [1.17] (rewritten by replacing natural logarithm by decimal logarithm), the coefficients A and B are then the solutions to the system:

$$\begin{cases} \log_{10} P_c = A - B / T_c \\ \log_{10}\left[P_c \cdot 10^{-(\omega+1)}\right] = A - B / (0.7T_c) \end{cases} \qquad [1.20]$$

After resolution of system [1.20], we obtain expressions for the coefficients A and B that, when reinserted into equation [1.17], lead to:

$$P^{sat}(T) = P_c \cdot 10^{\frac{7}{3}(\omega+1)\left[1-\frac{(T_c/K)}{(T/K)}\right]} \qquad [1.21]$$

NOTE.– A relationship $P^{sat}(T)$ allows the calculation of saturated vapor pressure at a fixed temperature. The **boiling point temperature at a fixed pressure** is obtained by simply inverting the relationship $P^{sat}(T)$. For example, if we accept that the saturated vapor pressure is given by equation [1.17], the boiling point temperature then becomes:

$$\ln P^{sat}(T) = A - \frac{B}{(T/K)} \Leftrightarrow \left[T_{bp}(P)/K\right] = \frac{B}{A - \ln P}$$

If we now accept that the saturated vapor pressure is indeed produced by equation [1.21], then the boiling point at fixed pressure is given by:

$$P^{sat}(T) = P_c \cdot 10^{\frac{7}{3}(\omega+1)\left[1-\frac{(T_c/K)}{(T/K)}\right]} \Leftrightarrow \left[T_{bp}(P)/K\right] = \frac{(T_c/K)}{1 - \frac{3}{7(\omega+1)}\log_{10}\left(\frac{P}{P_c}\right)}$$

In order to describe the entire temperature range $\left[T_{triple} ; T_c\right]$ with a better precision than that provided by the relationship [1.17], it is important to use a more sophisticated expression involving three or four adjustable parameters. In order to obtain such expressions, the coefficient B should be no longer considered as a constant but as a polynomial of the temperature. For example, if we suppose that B takes the form:

$$B(T) = \alpha + \beta T + \gamma T^3 \qquad [1.22]$$

integration of the relationship [1.16] with respect to the variable T leads to:

$$\ln P^{sat}(T) = A' - \frac{B'}{(T/K)} + C' \ln(T/K) + D' \cdot (T/K)^2 \qquad [1.23]$$

where A', B', C' and D' are constant (adjustable parameters). This type of correlation is present in the majority of commercial process simulators.

1.3.2. *Empirical expressions (Antoine, Wagner, Frost–Kalkwarf)*

The Antoine equation is written as:

$$\ln P^{sat}(T) = A_A - \frac{B_A}{T + C_A}, \text{ valid over an interval } \left[T_{\min} ; T_{\max}\right] \qquad [1.24]$$

where A_A, B_A and C_A are three adjustable parameters that must nevertheless comply with certain constraints, for example, $B_A > 0$ (these constraints are later justified in the section "Focus on the Antoine equation").

The Wagner equation is written as:

$$\begin{cases} \ln\left[P^{sat}(T) / P_c\right] = \dfrac{A_W \cdot X + B_W \cdot X^{1,5} + C_W \cdot X^3 + D_W \cdot X^6}{1 - X} \\ X = 1 - (T / K) / (T_c / K) \end{cases} \qquad [1.25]$$

In the same way as the integrated form of the Clapeyron equation with four parameters, this expression allows a very precise representation of the saturated vapor pressures; this correlation can be used on the condition that sufficient experimental data are available for the estimation of the four adjustable parameters.

From a more anecdotal viewpoint, the Frost–Kalkwarf equation has the feature to be an implicit equation in P^{sat}. It is written as:

$$\ln P^{sat} = A_{FK} - \frac{B_{FK}}{(T / K)} + C_{FK} \cdot \ln(T / K) + D_{FK} \cdot \frac{P^{sat}}{(T / K)^2} \qquad [1.26]$$

and therefore needs to be solved by an appropriate numerical method in order to deduce P^{sat} at a fixed temperature.

1.3.3. *Taking things further: constraints on the curvature of the graphical representation of ln P^{sat} versus 1/T*

Figure 1.8 illustrates a quasi-systematic experimental observation verified on all pure substances: the curve $B(T) = -d \ln P^{sat} / d(\frac{1}{T})$ presents a minimum at a temperature generally included between T_{bp}° and T_c. The existence of this minimum induces a geometrical particularity on the $P^{sat}(T)$ curve that we will now discuss. Effectively, from equation [1.16], we deduce:

$$\frac{d^2 \ln P^{sat}}{d(1/T)^2} = -\frac{dB}{d(1/T)} = T^2 \frac{dB}{dT} \qquad [1.27]$$

Consequently, the presence of a minimum on the curve $B(T)$ (characterized by $dB/dT = 0$) necessarily leads to the presence of an inflection point on the curve $\ln P^{sat}$ vs.$1/T$ (characterized by $d^2 \ln P^{sat}/d(1/T)^2 = 0$).

It follows that the existence of a minimum on the $B(T)$ curve indicates that, strictly speaking, the curve $\ln P^{sat}$ vs.$1/T$ is not exactly a straight line but a convex curve with low values of $1/T$ and then a concave curve for the rest of the domain. These conclusions are summarized in Figure 1.8.

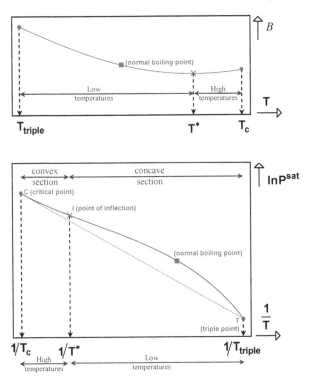

Figure 1.8. *Graphical illustration of the fact that a minimum on the B(T) curve (top) induces the presence of an inflection point on the curve* $\ln P^{sat}$ *versus 1/T (bottom)*

Strictly speaking, a correlation $P^{sat}(T)$ should be able to reproduce an inflection point in the plane $\ln P^{sat}$ vs. $1/T$. It is obvious that an affine correlation of $\ln P^{sat}$ vs. $1/T$ such as that provided by equation [1.17] is not able to reproduce an inflection point and induces differences between the experimental behavior and the behavior predicted by the correlation. How can the coefficients A and B in equation [1.17] be determined so as to minimize these differences? Let us look at some examples:

– if we select two points at low temperature ($T < T^*$, where T^* denotes the temperature of the experimental inflection point), in other words in the concave part of the curve $\ln P^{sat}$ vs. $1/T$, the coefficients A and B would lead to significant deviations for P^{sat} at high temperatures, as illustrated in Figure 1.9. This is typically what would be obtained when fitting A and B to the normal boiling point temperature and triple-point coordinates $\left(T_{triple} ; P_{triple}\right)$;

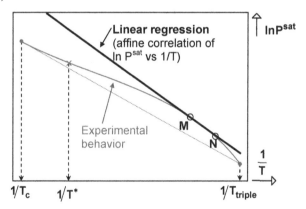

Figure 1.9. *Deviations induced at high temperatures between the experimental values of P^{sat} and those calculated using equation [1.17] when two points at low temperature (M and N) are considered to determine the coefficients A and B of the equation [1.17]*

– similarly, if two points at high temperature are selected ($T > T^*$), in other words, in the convex part of the $\ln P^{sat}$ vs. $1/T$ curve, the coefficients A and B would lead to significant deviations of P^{sat} at low temperature, as illustrated in Figure 1.10.

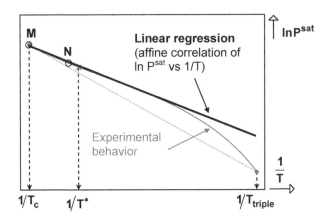

Figure 1.10. *Deviations induced at low temperature between the experimental values of P^{sat} and those calculated using equation [1.17] when two points at high temperature (M and N) are considered in order to determine the coefficients A and B of the equation [1.17]. For a color version of the figure, see www.iste.co.uk/ jaubert/thermodynamic.zip*

1.3.4. *Focus on the Antoine equation*

The Antoine equation, given by formula [1.24], is an empirical correlation that is often appreciated by engineers due to its simplicity and flexibility – which involves three adjustable parameters known as Antoine parameters – as well as due to the availability of Antoine parameters in the literature.

When deriving equation [1.24] with respect to the variable $1/T$, we obtain:

$$\frac{d \ln P^{sat}}{d\left(\frac{1}{T}\right)} = \frac{-B_A T^2}{\left(T + C_A\right)^2} \qquad [1.28]$$

Yet, the Clapeyron equation [1.16] stipulates that $\dfrac{d \ln P^{sat}}{d(1/T)}$ is equal to the

relationship $-\dfrac{1}{R} \times \dfrac{\Delta_{vap} H(T)}{\Delta_{vap} Z(T)}$, where $\Delta_{vap} H$ and $\Delta_{vap} Z$ are always positive

quantities, and $\dfrac{d \ln P^{sat}}{d(1/T)}$ must therefore be negative. Consequently, in order to comply with this constraint, **the coefficient B_A of the Antoine equation must be chosen to be positive** in order to make the ratio $-B_A T^2 / (T + C_A)^2$ positive.

Let us now ask ourselves about the ability of the Antoine equation to reproduce a point of inflection on the $\ln P^{sat}$ vs. $1/T$ curve and thus the concave part (such that $\dfrac{d^2 \ln P^{sat}}{d(1/T)^2} < 0$) and convex part (such that $\dfrac{d^2 \ln P^{sat}}{d(1/T)^2} > 0$). To answer this question, we express the derivative $\dfrac{d^2 \ln P^{sat}}{d(1/T)^2}$.

$$\frac{d^2 \ln P^{sat}}{d(1/T)^2} = \frac{2 \overbrace{B_A T^3 C_A}^{>0}}{(T + C_A)^3} = \frac{2 \overbrace{B_A T^3}^{>0}}{(T + C_A)^2} \frac{C_A}{(T + C_A)} \qquad [1.29]$$

Equation [1.29] indicates that $\dfrac{d^2 \ln P^{sat}}{d(1/T)^2}$ is always of the same sign as $\dfrac{C_A}{(T + C_A)}$. The condition for obtaining a point of inflection (such as $\dfrac{d^2 \ln P^{sat}}{d(1/T)^2} = 0$) at a discrete temperature T^* will therefore not be fulfilled by the Antoine equation. The Antoine equation stipulates that the curve $\ln P^{sat}$ vs. $1/T$ will always have the same curvature (either totally convex or totally concave):

– If $C_A > 0$, then $\dfrac{C_A}{(T + C_A)} > 0$, and consequently $\dfrac{d^2 \ln P^{sat}}{d(1/T)^2} > 0$. The curve $\ln P^{sat}$ vs. $1/T$ will then be convex over the entire temperature domain (whereas experimentally, this is only the case at high temperatures).

– If $C_A < 0$:

- either $\dfrac{C_A}{(T + C_A)} > 0 \Leftrightarrow T < -C_A$, and in this case: $\dfrac{d^2 \ln P^{sat}}{d(1/T)^2} > 0$, which means that the curve $\ln P^{sat}$ vs. $1/T$ is convex over the entire temperature domain (verified experimentally at high temperatures),

- or $\dfrac{C_A}{(T+C_A)}<0 \Leftrightarrow T>-C_A$, and in this case: $\dfrac{d^2 \ln P^{sat}}{d(1/T)^2}<0$, which

means that the curve $\ln P^{sat}$ vs. $1/T$ is concave over the entire temperature domain (verified experimentally at low temperatures).

In summary: either the Antoine equation is used to reproduce the **low temperature domain** (close to the triple point), and in this case, the constraint $\forall T : -T < C_A < 0$, in other words: $-T_{min} < C_A < 0$, must be imposed, where T_{min} indicates the lower limit of the temperature domain for the correlation **(this is often the selected solution)**, or the Antoine equation is used to reproduce the **high-pressure** domain (close to the critical point), and in this case, the constraint $\forall T : \begin{cases} C_A > 0 \\ \text{or } C_A < -T < 0 \end{cases}$, in other words: $\begin{cases} C_A > 0 \\ \text{or } C_A < -T_{max} < 0 \end{cases}$, must be imposed.

1.4. Equations that can be used to correlate the molar volumes of pure liquids

1.4.1. *Correlations that can be used for molar volumes of saturated liquids*

In this section, the description of the molar volumes of saturated liquids $v_L^{sat}(T)$ is discussed. In the pressure–molar volume plane, these are the molar volumes associated with the bubble point curve, as illustrated in Figure 1.11.

In practice, the molar volumes of saturated liquids are calculated with excellent accuracy (of the order of 1%) using the **Rackett equation**:

$$v_L^{sat}(T) = \frac{R \cdot T_c}{P_c} \times Z_{RA}^{\left[1+(1-T/T_c)^{2/7}\right]}$$ [1.30]

This correlation features the reduced temperature $T_r = T / T_c$ of the pure substance in addition to Z_{RA}, the *Rackett compressibility factor*, tabulated for many compounds by Spencer and Danner[3]. It is frequently implemented in commercial process simulators.

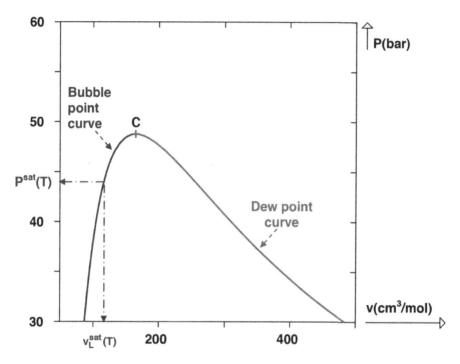

Figure 1.11. *Representation of the bubble points, whose coordinates are (v_L^{sat} (T) , P^{sat}(T)) in the P-v plane*

In order to increase the correlation accuracy, it is important, when Z_{RA} and an experimental point $\left(T_{exp}, v_{exp}\right)$ are known, to estimate the molar volume of the saturated liquid using:

$$v_L^{sat}\left(T\right) = v_{exp} \times Z_{RA}^{\left[\left(1-T/T_c\right)^{2/7} - \left(1-T_{exp}/T_c\right)^{2/7}\right]}$$ [1.31]

In process simulators, **the Costald equation** is also frequently implemented. It only requires knowledge of the acentric factor ω to be used, which is of the following form:

---

3 Spencer and Danner, *J. Chem. Eng. Data*, 17(2), 236–241, 1972.

$$\begin{cases} v_L^{sat}(T) = v^* v_r^{(0)} \left[1 - \omega v_r^{(\delta)} \right] \\ v^* = \text{ajustable parameter (can be replaced favorably} \\ \text{by } v_c \text{ in the absence of usable data)} \end{cases} \qquad [1.32]$$

The terms $v_r^{(0)}$ and $v_r^{(\delta)}$ are expressed by:

$$\begin{cases} v_r^{(0)} = 1 + a(1-T_r)^{1/3} + b(1-T_r)^{2/3} + c(1-T_r) + d(1-T_r)^{4/3} \\ a = -1.52816; b = 1.43907; c = -0.81446; d = 0.190454 \\ \text{valid for } T_r \in [0.25 ; 0.95] \end{cases}$$

$$\begin{cases} v_r^{(\delta)} = \dfrac{e + fT_r + gT_r^2 + hT_r^3}{T_r - 1.00001} \\ e = -0.296123; f = 0.386914; g = -0.0427458; h = -0.0480645 \\ \text{valid for } T_r \in [0.25 ; 1.00] \end{cases} \qquad [1.33]$$

Certain commercial databases such as the "DIPPR[4] Database" also frequently use highly flexible correlations that contain a large number of adjustable parameters to correlate their data. For example:

$$v_L^{sat}(T) = \frac{C_1^{\left[1 + (1 - T/C_2)^{C_3} \right]}}{C_4} \qquad [1.34]$$

where the coefficients C_i are fit to experimental data.

1.4.2. Correlations that can be used for molar volumes of subcooled liquids

In light of the Gibbs phase rule, the variance of a pure single-phase substance is equal to 2, and consequently, the molar volumes of subcooled liquids depend on two intensive variables of the liquid phase. Generally, the temperature and the pressure are chosen. To evaluate these data, an equation of state can be used (these models are described in the next chapter).

---

4 Design Institute for Physical Properties.

However, under moderate pressure (in general, the range $P < 30$ bar is considered), the *incompressible liquid* **assumption** leads to an accurate estimation of liquid molar volumes.

This assumption means that the influence of the pressure on the liquid molar volumes can be neglected. We therefore have:

$$v_L(T,P) \approx v_L(T,P^{sat}(T)) = v_L^{sat}(T) \qquad [1.35]$$

The correlations for the molar volumes of saturated liquids can then be used to estimate the molar volumes of subcooled liquids.

1.5. Equations that can be used to correlate the molar volumes of saturated gases

When the saturated vapor pressure is known and sufficiently low (typically <5 bar), the perfect gas law is used to estimate the molar volume of a saturated gas:

$$v_G^{sat}(T) = RT / P^{sat}(T) \qquad [1.36]$$

In the other cases, the **law of rectilinear diameter** is normally used; it comes from an experimental observation: the average value of the molar densities $(\rho = 1/v)$ of the saturated liquid and of the saturated gas is approximately a linear function of the temperature:

$$0.5\left(\frac{1}{v_G^{sat}(T)} + \frac{1}{v_L^{sat}(T)}\right) = \alpha T + \beta \qquad [1.37]$$

where α and β are parameters that can be fit to experimental data. Thus, with known α, β and a correlation for $v_L^{sat}(T)$ (e.g. the Rackett equation or the Costald equation), it becomes possible to access $v_G^{sat}(T)$.

If, in addition, the critical molar volume and the critical temperature are known, the law of rectilinear diameter is written as:

$$0.5\left(\frac{v_c}{v_G^{sat}(T)} + \frac{v_c}{v_L^{sat}(T)}\right) = 1 + \alpha\left(1 - \frac{T}{T_c}\right) \qquad [1.38]$$

The constant α, which should be regressed on experimental data, is often close to 0.75.

1.6. Equations that can be used to correlate the enthalpies of vaporization

1.6.1. *Use of the Clapeyron equation*

From the Clapeyron equation [1.12], we deduce:

$$\Delta_{vap}H(T) = T \times \Delta_{vap}V(T) \times P^{sat}(T) \times \frac{d\ln P^{sat}}{dT}(T) \qquad [1.39]$$

Let us assume that we have a suitable correlation for $P^{sat}(T)$; it then becomes possible, by simple derivation, to calculate $\dfrac{d\ln P^{sat}}{dT}$, as illustrated in Table 1.1.

Correlation $P^{sat}(T)$	Equation	$d\ln P^{sat}/dT$
Antoine	[1.24]	$B_{A} / \left(T + C_{A} \right)^{2}$
integrated Clapeyron	[1.17]	B / T^{2}
integrated extended Clapeyron	[1.23]	$\dfrac{B'}{T^{2}} + \dfrac{C'}{T} + 2D' \cdot T$

Table 1.1. *Expressions of the derivatives of* ln P^{sat} *with respect to T according to various correlations (Antoine, integrated form of the Clapeyron equation, integrated extended form of the Clapeyron equation)*

For an estimation of the term $\Delta_{vap}V(T) = v_{G}^{sat}(T) - v_{L}^{sat}(T)$, various strategies can be used:

– **Case 1:** if $P < 5$ bar and if $T \ll T_{c}$, the saturated gas can be assumed perfect and the molar volume of the saturated liquid can be considered negligible in comparison to the saturated gas. It then follows that $\Delta_{vap}V(T) \approx v_{G}^{sat}(T) = RT / P^{sat}(T)$, and consequently:

$$\Delta_{vap}H(T) \approx RT^2 \frac{d\ln P^{sat}}{dT} \tag{1.40}$$

– **Case 2:** if acceptable correlations for $v_G^{sat}(T)$ (e.g. the law of rectilinear diameter) and $v_L^{sat}(T)$ (e.g. the Rackett equation or the Costald equation) are available, $\Delta_{vap}V(T)$ can be evaluated simply and equation [1.39] can be used as is.

1.6.2. *Expression involving many adjustable parameters*

As previously mentioned, certain commercial databases use correlations that involve a large number of adjustable parameters to be fit to experimental data. For example:

$$\begin{cases} \Delta_{vap}H(T_r) = C_1\left(1-T_r\right)^{C_2+C_3T_r+C_4T_r^2+C_5T_r^3} \\ \text{with: } T_r = T / T_c \end{cases} \tag{1.41}$$

where the coefficients C_i are constant parameters. Note that the proposed equation complies with the constraint at the critical point: $\Delta_{vap}H(T_r=1)=0$.

1.7. Equations used to correlate the heat capacity at constant molar pressure, molar enthalpy or molar entropy of an incompressible liquid

The *incompressible liquid* assumption allows us to neglect the influence of the pressure on the value of an intensive liquid property. Thus:

$$\begin{cases} h_L(T,\cancel{P}) \approx h_L^{sat}(T) \\ c_{P,L}(T,\cancel{P}) \approx c_{P,L}^{sat}(T) \end{cases} \tag{1.42}$$

Let us recall the definition of the molar heat capacity at constant pressure:

$$c_P = \left(\frac{\partial h}{\partial T}\right)_P \tag{1.43}$$

In the case of an incompressible liquid, we have:

$$c_{P,L} = \frac{dh_L}{dT} \qquad [1.44]$$

The molar heat capacities of saturated liquids can be considered constant over a restricted temperature range or as functions of the temperature over large temperature ranges. They are frequently expressed using polynomials of the temperature, for example:

$$c_{P,L}^{sat}(T) = C_0 + C_1 T + C_2 T^2 + C_3 T^3 + C_4 T^4 \qquad [1.45]$$

The coefficients C_i are constants that are fit to experimental data.

On the condition that an acceptable correlation for $c_{P,L}^{sat}(T)$ is known, it becomes possible to express the change of the molar enthalpy of an incompressible liquid with respect to temperature by integrating the relationship [1.44]:

$$h_L(T) - h_L(T_0) = \int_{T_0}^{T} c_{P,L}^{sat}(T) \cdot dT \qquad [1.46]$$

Insofar as the energy balances only reveal differences in molar enthalpies, the latter can be defined to within one constant. In order to express the molar enthalpies, we then resort to the choice of a **reference state**. This consists of arbitrarily supposing that the molar enthalpy of a pure liquid at a reference temperature T_0 is zero: $h_L(T_0) = 0$. We therefore obtain:

$$h_L(T) = \int_{T_0}^{T} c_{P,L}^{sat}(T) \cdot dT \qquad [1.47]$$

Other choices of reference states would be possible, but they will not be discussed in the context of this book, which focuses on the estimation of the properties of matter and not on the various ways of writing energy balances.

Let us now consider molar entropy. In the case of liquid pure substances, the molar entropy is linked to the molar enthalpy by the thermodynamic identity:

$$dh_L = Tds_L + v_L dP \underset{\substack{incompressible \\ liquid}}{\approx} Tds_L \qquad [1.48]$$

Combining equations [1.44] and [1.48], it follows that:

$$ds_L = c_{P,L} \frac{dT}{T} \qquad [1.49]$$

which, in integrated form, is written as:

$$s_L(T) - s_L(T_0) = \int_{T_0}^{T} c_{P,L}^{sat}(T) \cdot \frac{dT}{T} \qquad [1.50]$$

Following the same principle as for the enthalpies, insofar as the entropy balances only involve the differences of molar entropies of the components, we can express the molar entropy of a pure incompressible liquid to within one constant. Choosing arbitrarily $s_L(T_0) = 0$, we obtain

$$s_L(T) = \int_{T_0}^{T} c_{P,L}^{sat}(T) \cdot \frac{dT}{T} \qquad [1.51]$$

1.8. Equations that can be used to correlate the molar heat capacity at constant pressure, molar enthalpy or molar entropy of a perfect gas

In compliance with the experimental observations, statistical thermodynamics allows us to prove that the molar heat capacity at constant pressure and the molar enthalpy of a pure perfect gas only depend on the temperature.

– In the case of a **single-atom perfect gas** (He, Ne, Ar, single-atom metal vapors), the gas kinetic theory allows us to demonstrate that the molar heat capacity at constant pressure is constant for the entire range of temperatures and is equal to:

$$c_P^{\bullet} = \frac{5}{2} R \qquad [1.52]$$

where $R = 8.3144 \ \text{J} \cdot \text{mol}^{-1} \cdot \text{K}^{-1}$ is the gas constant.

– In the case of a **diatomic perfect gas** (O_2, N_2, H_2, etc.), kinetic theory of gases allows us to show that the molar heat capacity at constant pressure is constant over a wide range of medial temperatures (covering approximately the range 100–1000 K):

$$c_P^\bullet = \frac{7}{2}R \qquad\qquad [1.53]$$

At very low temperatures, we find $c_P^\bullet = \frac{5}{2}R$, whereas at very high temperatures, we find $c_P^\bullet = \frac{9}{2}R$.

– In other cases, we resort to semi-empirical formulae (which sometimes combine theoretical or semi-theoretical expressions and parameters fit to experimental data). For example:

$$c_P^\bullet (T) = C_1 + C_2 \left[\frac{C_3 / T}{\sinh\left(C_3 / T\right)} \right]^2 + C_4 \left[\frac{C_5 / T}{\cosh\left(C_5 / T\right)} \right]^2 \qquad [1.54]$$

In the same way as the molar enthalpy of an incompressible liquid, the molar enthalpy of a perfect gas can be deduced from the relationship:

$$h^\bullet(T) - h^\bullet(T_0) = \int_{T_0}^{T} c_P^\bullet(T) \cdot dT \qquad\qquad [1.55]$$

If the reference state that was previously chosen in the paragraph referring to incompressible liquids is conserved, i.e. $h_L(T_0) = 0$, it follows that:

$$h^\bullet(T) = h^\bullet(T_0) + \int_{T_0}^{T} c_P^\bullet(T) \cdot dT$$

$$= h^\bullet(T_0) - h_L(T_0) + \int_{T_0}^{T} c_P^\bullet(T) \cdot dT \qquad [1.56]$$

Noting that $h^{\bullet}(T_0) - h_L(T_0)$ identifies with the enthalpy of vaporization of a pure substance at T_0 (on the condition that at this temperature, the saturated gas can be considered perfect and that the saturated liquid can be assumed incompressible), it follows that:

$$h^{\bullet}(T) \approx \Delta_{vap}H(T_0) + \int_{T_0}^{T} c_P^{\bullet}(T) \cdot dT \qquad [1.57]$$

The molar entropy of a perfect gas is related to its molar enthalpy by:

$$\underbrace{dh^{\bullet}}_{c_P^{\bullet}dT} = Tds^{\bullet} + v^{\bullet}dP = Tds^{\bullet} + \frac{RT}{P}dP \qquad [1.58]$$

Therefore:

$$ds^{\bullet} = c_P^{\bullet}\frac{dT}{T} - R\frac{dP}{P} \qquad [1.59]$$

It follows that:

$$s^{\bullet}(T,P) - s^{\bullet}(T_0,P_0) = \int_{T_0}^{T} c_P^{\bullet}(T) \cdot \frac{dT}{T} - R\ln\left(\frac{P}{P_0}\right) \qquad [1.60]$$

In choosing $P_0 = P^{sat}(T_0)$ and remembering that the chosen reference state is such that $s_L(T_0) = 0$ (molar entropy of the incompressible liquid at T_0), we obtain the molar entropy of vaporization at $T_0 : s^{\bullet}(T_0, P^{sat}(T_0)) - s_L(T_0)$ and thus:

$$s^{\bullet}(T,P) = \Delta_{vap}S(T_0) + \int_{T_0}^{T} c_P^{\bullet}(T) \cdot \frac{dT}{T} - R\ln\left[\frac{P}{P^{sat}(T_0)}\right] \qquad [1.61]$$

1.9. Density of a pure substance

The density of a pure substance is related to its molar volume by:

$$\bar{\rho}(T,P) = \frac{M}{v(T,P)} \qquad [1.62]$$

where M denotes the molar mass of the pure substance and v denotes its molar volume. All the expressions that have previously been mentioned in order to estimate v are therefore usable to access the volumetric mass.

1.10. Prediction of the thermodynamic properties of pure substances

A method for predicting the properties of pure substances is said to be **predictive** if its application only requires knowledge of structural information on the molecule considered and does not require prior knowledge of adjustable parameters that would be specific to it.

These methods are often based on the concept of **group contributions**: each molecule is seen as an assembly of elementary molecular groups. An example of decomposition of propan-1-ol into elementary groups is proposed in Figure 1.12.

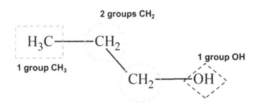

Figure 1.12. *Decomposition of propan-1-ol into elementary groups*

We are first interested in any intensive property Y, independent of the temperature. In order to estimate the property Y_k of a molecule k by contribution of groups, we postulate that in the same way as the molar mass, Y_k can be seen as the sum of the **group occurrences** n_i^k (in other words, of the number of times that the groups i appear in the molecule k) weighted using coefficients y_i expressing the **contribution from the group i to the value of the property Y_k**.

$$Y_k = \overset{\underset{\text{Number of groups}}{\text{defined by the method}}}{\underset{i=1}{\overset{k}{\sum}}} n_i^k \cdot y_i \qquad\qquad [1.63]$$

For example, in the case of propan-1-ol, we will write $Y_{\text{propan-1-ol}} = y_{OH} + 2y_{CH_2} + y_{CH_3}$.

In practice, equation [1.63] is rarely used as is, and has to be adapted. It is not unusual to replace the property Y by a function of Y denoted as $f(Y)$. It is thus assumed that $f(Y)$ is compatible with the group-contribution concept; in other words, it can be expressed as a linear function of the group occurrences:

$$f(Y_k) = \overset{\underset{\text{Number of groups}}{\text{defined by the method}}}{\underset{i=1}{\overset{k}{\sum}}} n_i^k \cdot y_i \quad with: \begin{cases} f(Y) = \alpha Y + \beta \\ f(Y) = \ln(Y + \beta) \\ f(Y) = Y^\alpha \\ \vdots \end{cases} \qquad [1.64]$$

As an example, the Joback–Reid method assumes that the normal boiling point temperature is given by:

$$T_{bp}^\circ - 198,2 = \overset{\underset{\text{Number of groups}}{\text{defined by the method}}}{\underset{i=1}{\overset{k}{\sum}}} n_i^k \cdot tb_i \qquad\qquad [1.65]$$

The coefficients tb_i are tabulated in the original article by Joback and Reid for a large number of groups.

Many predictive methods for estimating the properties of pure substances are described in the literature. Among them, we note mainly:

– Joback–Reid method[5];

– Constantinou–Gani method[6];

– Marrero–Gani method[7].

5 K.G. Joback, R.C. Reid, *Chem. Eng. Commun.*, 57, pp. 233–243, 1987.
6 L. Constantinou, R. Gani, *AIChE J.*, 40(10), pp. 1697–1710, 1994.
7 J. Marrero, R. Gani, *Fluid Phase Equilibria*, 183–184, pp. 183–208, 2001.

Estimation of Thermodynamic Properties of Pure Substances Using an Equation of State: Overview of Available Models and Calculation Procedures

2.1. Volumetric equation of state: a definition

Experimental postulate: for uniform closed systems of constant composition, which only receive volumetric work from pressure forces, the change of state functions can be represented using two independent state variables. **We say that uniform closed systems of constant composition are divariant.**

Consequence: there are mathematical equations that relate each state property to two others, for example, $P(T,v)$ or $U(T,P)$. Such models are called *equations of state*. In particular, the mathematical relationship that links the variables P, v, T to each other is called a **volumetric equation of state**, for example, $P(T,v) = RT/v$.

2.2. General overview of the volumetric equations of state

There are two large categories of equations of state for pure substances:

Pressure-explicit equations of state: they are of the form $P = f(T,v)$ and can be used to model a gas phase and/or a liquid phase and/or a diphasic vapor–liquid equilibrium. The main types of pressure-explicit equations of state are:

– the truncated virial equation of state: $P(T,v) = RT / [v - B(T)]$ (applicable to a gas);

– the virial expansion of the pressure: $P(T,v) = RT / v + BRT / v^2 + CRT / v^3 + ...$ (applicable to a gas);

– the cubic equations of state (applicable to a liquid, a gas or a system in vapor–liquid equilibrium);

– the SAFT[1] equations of state (applicable to a liquid, a gas or a system in vapor–liquid equilibrium);

– the equations of state that are specific to particular pure substances (or mixtures). They have multiple parameters and are frequently based on expansions of the Helmholtz energy (Span–Wagner, GERG, etc.); they are applicable to a liquid, a gas or a system in vapor–liquid equilibrium.

Volume-explicit equations of state: they are of the form $v = f(T,P)$. They can only be used to model a single-phase state (gas in practice). Because at fixed temperature and pressure these models only predict a single molar volume, they cannot predict phase equilibrium.

2.3. Presentation of usual volumetric equations of state

2.3.1. *Virial expansion*

The quantity $z = Pv / (RT)$ is known as the ***compressibility factor*** (it is pointed out that this quantity is equal to 1 for a perfect gas).

There are two types of virial expansions:

– Expression of the compressibility factor as a power series in the molar density variable $1/v$:

$$z(T,v) = \frac{Pv}{RT} = 1 + \frac{B(T)}{v} + \frac{C(T)}{v^2} + \frac{D(T)}{v^3} + ...$$ [2.1]

---

1 SAFT means "statistical associating fluid theory".

where $B(T)$, $C(T)$, $D(T)$, etc. are called **virial coefficients**. The expression for the equation of state is then:

$$P(T,v) = \frac{RT}{v} + \frac{RTB(T)}{v^2} + \frac{RTC(T)}{v^3} + \frac{RTD(T)}{v^4} + \dots \qquad [2.2]$$

Theoretically applicable to a fluid (in a one-phase gas or liquid state or in vapor–liquid equilibrium), this type of development is in practice only used to describe gas systems. The number of terms that need to be considered to properly represent the properties of the gas increases with increasing pressure.

– Expression of the compressibility factor as a power series in pressure P:

$$z(T,P) = \frac{Pv}{RT} = 1 + B'(T) \cdot P + C'(T) \cdot P^2 + D'(T) \cdot P^3 + \dots \qquad [2.3]$$

As with the virial coefficients, $B'(T)$, $C'(T)$, $D'(T)$, etc. are functions only of the temperature. The expression for the equation of state is:

$$v(T,P) = \frac{RT}{P} + RTB'(T) + RTC'(T) \cdot P + RTD'(T) \cdot P^2 + \dots \qquad [2.4]$$

This type of equation of state is only applicable to gas systems;

– Link between the two types of power series:

Substituting the variable pressure in the expansion [2.3] by its expression from the expansion [2.2], we have:

$$
\begin{aligned}
z &= 1 + B'(T) \cdot P + C'(T) \cdot P^2 + D'(T) \cdot P^3 + \dots \\
&= 1 + B'(T) \cdot \left[\frac{RT}{v} + \frac{RTB(T)}{v^2} + \dots \right] + C'(T) \cdot \left[\frac{RT}{v} + \frac{RTB(T)}{v^2} + \dots \right]^2 \\
&\quad + D'(T) \cdot \left[\frac{RT}{v} + \frac{RTB(T)}{v^2} + \dots \right]^3 + \dots \\
&= 1 + \frac{(B'RT)}{v} + \frac{(B'BRT + C'R^2T^2)}{v^2} + \frac{(B'CRT + 2C'BR^2T^2 + D'R^3T^3)}{v^3} + \dots
\end{aligned}
$$

$$[2.5]$$

By identification with the terms of equation [2.1], we obtain:

$$\begin{cases} B' = \dfrac{B}{RT} \\[2mm] C' = \dfrac{C - B^2}{R^2 T^2} \\[2mm] D' = \dfrac{D - 3BC + 2B^3}{R^3 T^3} \\[2mm] \dots \end{cases} \qquad [2.6]$$

– Truncated virial equation:

According to equations [2.1], [2.3] and [2.6], the following two virial expansions are strictly equivalent when an infinite number of terms are considered:

$$\begin{cases} z(T,v) = 1 + \dfrac{B}{v} + \dfrac{C}{v^2} + \dfrac{D}{v^3} + \dots \\[2mm] z(T,P) = 1 + \dfrac{B}{RT} P + \dfrac{C - B^2}{R^2 T^2} P^2 + \dfrac{D - 3BC + 2B^3}{R^3 T^3} P^3 + \dots \end{cases} \qquad [2.7]$$

It must be noted, however, that with the exception of the first, the terms are not identical two by two.

Under moderate pressure (typically $P < 20$ bar), it is possible to carry out the **slightly imperfect gases approximation**, considering only the first two terms for each development:

$$z(T,v) = 1 + \frac{B(T)}{v} \quad or \quad z(T,P) = 1 + \frac{B(T)P}{RT} \qquad [2.8]$$

Two different equations are obtained, but with equivalent accuracy. Often, the expression in variables (T,P) is preferred:

$$z(T,P) = 1 + \frac{B(T)P}{RT} \quad \Leftrightarrow \quad \begin{cases} v(T,P) = \dfrac{RT}{P} + B(T) \\[2mm] P(T,v) = \dfrac{RT}{v - B(T)} \end{cases} \qquad [2.9]$$

These expressions define the truncated virial equation.

– Estimate of the second virial coefficient $B(T)$:

This coefficient has the dimension of a molar volume. The graph in Figure 2.1 shows the general evolution of this property as a function of temperature.

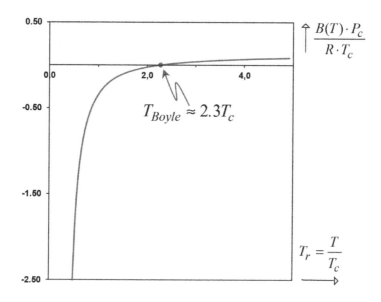

Figure 2.1. *General shape of the second virial coefficient as a function of temperature*

We note that second virial coefficients are increasing functions of temperature. At low temperatures, $B(T)$ is negative and the temperature that cancels the second virial coefficient is known as the Boyle temperature. Generally, it is between 2 and 2.7 times the critical temperature. Many correlations are available in the literature for an estimation of $B(T)$. As an example, the DIPPR databank reports the coefficients of these correlations for more than a thousand components. It should be noted that semi-predictive methods also exist (they require knowledge of an experimental datum to estimate $B(T)$). For example, the Tsonopoulos correlation expresses the dimensionless property $BP_c / (RT_c)$ as a universal function of the reduced temperature $T_r = T/T_c$ and the acentric factor ω (in application of the three-parameter corresponding-state law):

$$\frac{B(T)P_c}{RT_c} = f^{(0)}(T_r) + \omega f^{(1)}(T_r) + f^{(2)}(T_r)$$

with :
$$\begin{cases} f^{(0)}(T_r) = 0.1445 - \dfrac{0.0330}{T_r} - \dfrac{0.1385}{T_r^2} - \dfrac{0.0121}{T_r^3} - \dfrac{0.000607}{T_r^8} \\[3mm] f^{(1)}(T_r) = 0.0637 + \dfrac{0.331}{T_r^2} - \dfrac{0.423}{T_r^3} - \dfrac{0.008}{T_r^8} \\[3mm] f^{(2)}(T_r) = \dfrac{a}{T_r^6} - \dfrac{b}{T_r^8} \ \text{(for polar / associated molecules)} \end{cases}$$ [2.10]

Coefficients a and b are zero for compounds that are non-polar and non-associating (through hydrogen bonds) such as alkanes. In other cases (alcohols, amines, etc.), they must be known and are specific to the component under consideration.

2.3.2. Classic cubic equations of state

– Summary:

Such models can describe a liquid phase or a gas phase or a system in vapor–liquid equilibrium. The term "*cubic*" means that for a given temperature and pressure, the molar volume is solution of a third-degree equation. The main cubic equations of state are presented in Table 2.1.

More generally, all cubic equations of state can be written in the form:

$$P(T,v) = \frac{RT}{v-b} - \frac{a_c \cdot \alpha(T)}{(v-br_1)(v-br_2)} \ \text{defined for} \ \begin{cases} T > 0 \\ v > b \end{cases}$$

$$with : \begin{cases} \text{Van der Waals} : r_1 = r_2 = 0 \\ \text{SRK} : \qquad\qquad r_1 = -1 \ and \ r_2 = 0 \\ \text{PR} : \qquad\qquad r_1 = -1 - \sqrt{2} \ and \ r_2 = -1 + \sqrt{2} \end{cases}$$ [2.11]

The parameters a_c and b are given by the general expressions:

$$\begin{cases} a_c = \Omega_a R^2 T_{c,\exp}^2 / P_{c,\exp} \\ b = \Omega_b R T_{c,\exp} / P_{c,\exp} \end{cases} \qquad [2.12]$$

The parameters Ω_a and Ω_b are **universal constants** of the equation of state. In other words, for a given equation of state (characterized by fixed r_1 and r_2 values), they keep the same values for all components under consideration. Their expression – given by equation [2.13] – is generally determined in such a way that the temperature and the pressure of the critical point, as expressed by the equation of state, are exactly equal to their experimental values $T_{c,\exp}$ and $P_{c,\exp}$ (see hereafter).

Critical constraints $\Rightarrow$

$$\begin{cases} \Omega_a = \dfrac{\left(1 - \eta_c \cdot r_1\right)\left(1 - \eta_c \cdot r_2\right)\left[2 - \eta_c(r_1 + r_2)\right]}{\left(1 - \eta_c\right)\left[3 - \eta_c\left(1 + r_1 + r_2\right)\right]^2} \\[12pt] \Omega_b = \dfrac{\eta_c}{3 - \eta_c\left(1 + r_1 + r_2\right)} \\[12pt] \text{with: } \eta_c = \left[1 + \sqrt[3]{(1 - r_1)(1 - r_2)^2} + \sqrt[3]{(1 - r_2)(1 - r_1)^2}\right]^{-1} \end{cases} \qquad [2.13]$$

It should be noted that most cubic equations of state are parameterized in such a way as to reproduce experimental critical pressures and temperatures of pure substances. $T_{c,\exp}$ **and** $P_{c,\exp}$ **are therefore input parameters for most cubic equations of state applicable to pure substances.**

– Critical constraints:

We are forcing the critical isotherm ($T = T_{c,\exp}$) to exhibit an **inflection point with a horizontal tangent** in the (P, v) plane when the pressure is $P = P_{c,\exp}$ (as seen in the previous chapter, such an infection point characterizes a critical point). These constraints result in a system of three equations:

Cubic equation of state	Date of publication	Expression	
Van der Waals	1873	$P(T,v) = \dfrac{RT}{v-b} - \dfrac{a_c}{v^2}$ **Effectiveness of the model:** *qualitative:* ☺☺☺ *quantitative:* ☹☹	$\begin{cases} a_c = \dfrac{27}{64} \dfrac{R^2 T_{c,exp}^2}{P_{c,exp}} \\[2ex] b = \dfrac{1}{8} \dfrac{RT_{c,exp}}{P_{c,exp}} \end{cases}$
Soave–Redlich–Kwong (SRK)	1972	$P(T,v) = \dfrac{RT}{v-b} - \dfrac{a(T)}{v(v+b)}$ **Effectiveness of the model:** *qualitative:* ☺☺☺ *quantitative:* ☺ (clear improvement of saturated vapor pressures thanks to the introduction of the α-function)	$\begin{cases} T_r = T/T_{c,exp} \\ a(T_r) = a_c \cdot \alpha(T_r) \\ \alpha(T_r) = \\ \quad \left[1 + m\left(1 - \sqrt{T_r}\right)\right]^2 \\ m = 0.480 + 1,574\omega \\ \quad - 0.176\omega^2 \\ a_c = 0.42748 \dfrac{R^2 T_{c,exp}^2}{P_{c,exp}} \\ b = 0.08664 \dfrac{RT_{c,exp}}{P_{c,exp}} \end{cases}$
Peng–Robinson (PR)	1976	$P(T,v) = \dfrac{RT}{v-b}$ $\quad - \dfrac{a(T)}{v(v+b) + b(v-b)}$ **Effectiveness of the model:** *qualitative:* ☺☺☺ *quantitative:* ☺☺ (same quality of reproduction of saturated vapor pressures as SRK, clear improvement of the reproduction of the liquid densities in comparison to SRK)	$\begin{cases} T_r = T/T_{c,exp} \\ a(T_r) = a_c \cdot \alpha(T_r) \\ \alpha(T_r) = \\ \quad \left[1 + m\left(1 - \sqrt{T_r}\right)\right]^2 \\ m = 0.37464 + 1,54226\omega \\ \quad - 0.26992\omega^2 \\ a_c = 0.45724 \dfrac{R^2 T_{c,exp}^2}{P_{c,exp}} \\ b = 0.07780 \dfrac{RT_{c,exp}}{P_{c,exp}} \end{cases}$

Table 2.1. *The most common cubic equations of state*

$$\begin{cases} P_{c,\mathrm{exp}} = P(T_{c,\mathrm{exp}}, v_c) \\[2mm] \left(\dfrac{\partial P}{\partial v} \right)_T \Bigg|_{\substack{T=T_{c,\mathrm{exp}} \\ v=v_c}} = 0 \\[4mm] \left(\dfrac{\partial^2 P}{\partial v^2} \right)_T \Bigg|_{\substack{T=T_{c,\mathrm{exp}} \\ v=v_c}} = 0 \end{cases}$$ [2.14]

As previously explained, in the case of cubic equations of state, the critical constraints are generally used to determine expressions for the parameters a_c and b (and to deduce values of the parameters Ω_a and Ω_b from them). The Van der Waals equation is now chosen to illustrate the application of the critical constraints. The mathematical expression of the Van der Waals equation is given in Table 2.1 and leads to:

$$\begin{cases} \left(\dfrac{\partial P}{\partial v} \right)_T = \dfrac{-RT}{(v-b)^2} + \dfrac{2a_c}{v^3} \\[4mm] \left(\dfrac{\partial^2 P}{\partial v^2} \right)_T = \dfrac{2RT}{(v-b)^3} - \dfrac{6a_c}{v^4} \end{cases}$$ [2.15]

The critical constraints [2.14] combined with equation [2.15] give:

$$\begin{cases} \dfrac{RT_{c,\mathrm{exp}}}{(v_c-b)^2} = \dfrac{2a_c}{v_c^3} \\[4mm] \dfrac{RT_{c,\mathrm{exp}}}{(v_c-b)^3} = \dfrac{3a_c}{v_c^4} \\[4mm] P_{c,\mathrm{exp}} = \dfrac{RT_{c,\mathrm{exp}}}{v_c-b} - \dfrac{a_c}{v_c^2} \end{cases}$$ [2.16]

The system [2.16] involves three unknowns: a_c, b and v_c (the critical molar volume). Once resolved, the system leads to the following expressions:

$$\begin{cases} a_c = \dfrac{27}{64} \dfrac{R^2 T_{c,\exp}^2}{P_{c,\exp}} \\[2ex] b = \dfrac{1}{8} \dfrac{RT_{c,\exp}}{P_{c,\exp}} \\[2ex] v_c = \dfrac{3}{8} \dfrac{RT_{c,\exp}}{P_{c,\exp}} \quad \Leftrightarrow \quad z_c = \dfrac{P_{c,\exp} v_c}{RT_{c,\exp}} = \dfrac{3}{8} = 0.375 \end{cases} \qquad [2.17]$$

We thus find the values of Ω_a and Ω_b that are reported in Table 2.1. We should note that the Van der Waals equation predicts a universal critical molar compressibility factor (the same for all pure substances), which does not comply with experimental observation (in practice, z_c varies between 0.18 and 0.32 depending on the nature of the molecules; its distribution is centered on the average value 0.27).

– The importance of the correct choice of the α-function:

The main advantage of the cubic equations of state is their capacity to accurately correlate (with a few-percent deviation) saturated vapor pressures, vaporization enthalpies and heat capacities of pure substances. Compounds involving association (by hydrogen bonds) exhibit the highest deviations. The choice of the α-function is an essential factor to guarantee accuracy of the model. We have recently demonstrated that this function should verify the following properties:

– For all $T_r > 0$, $\alpha(T_r) > 0$ and α must be decreasing and convex (and ideally, its third derivative must be positive),

– $\alpha(T_r = 1) = 1$,

– $\lim\limits_{T_r \to +\infty} \alpha(T_r) = 0$.

These characteristics are illustrated in Figure 2.2. There are a very high number of α-functions[2]. Very few verify all the properties listed above. These functions can be mainly classified into two categories: **universal functions** and **specific functions**. **Universal functions** do not include parameters that need to be fitted to (thermodynamic) data and depend on an

---

2 J.O. Valderrama, *Ind. Eng. Chem. Res.*, 42, pp. 1603–1618, 2003.

experimental property, generally the acentric factor ω_{exp}. Their strength lies in their semi-predictive nature. Well known for its effectiveness, Soave's function is definitely the most frequently used:

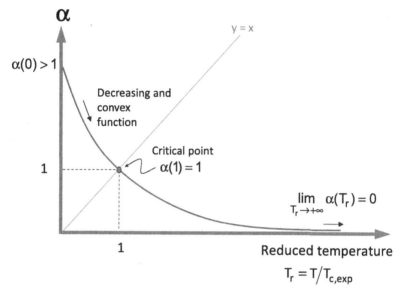

Figure 2.2. *General form and expected properties of the α function of cubic equations of state*

$$\alpha_{Soave}(T_r,\omega) = \left[1 + m(\omega) \cdot \left(1 - \sqrt{T_r}\right)\right]^2 \qquad [2.18]$$

The expression $m(\omega)$ depends on the chosen equation of state (see Table 2.1).

The universal *generalized Twu* function can also be mentioned:

$$\begin{cases} \alpha_{Gen.\,Twu}(T_r,\omega) = \alpha^{(0)}(T_r) + \omega \cdot \left[\alpha^{(1)}(T_r) - \alpha^{(0)}(T_r)\right] \\ \alpha^{(i)} = T_r^{N_i(M_i-1)} e^{L_i\left(1-T_r^{M_iN_i}\right)} \end{cases} \qquad [2.19]$$

The parameters of the generalized Twu function are provided in Table 2.2. We should note that it is a piecewise α function with an expression for $T_r \leq 1$ and another one for $T_r > 1$. Although it is frequent to use piecewise α

functions (Boston–Mathias, Stryjek–Vera, etc.), this type of function is not satisfactory because near the critical temperature, it induces discontinuities for so-called derivative properties such as enthalpy and heat capacities.

Parameters of the alpha function	$T \leq T_c$		$T > T_c$	
	$\alpha^{(0)}$	$\alpha^{(1)}$	$\alpha^{(0)}$	$\alpha^{(1)}$
Equation of state	PR:			
L	0.125283	0.511614	0.401219	0.024955
M	0.911807	0.784054	4.963070	1.248089
N	1.948150	2.812520	−0.200000	−8.000000
Equation of state	SRK:			
L	0.141599	0.500315	0.441411	0.032580
M	0.919422	0.799457	6.500018	1.289098
N	2.496441	3.291790	−0.200000	−8.000000

Table 2.2. *Parameters of the generalized Twu function*

The **specific functions** involve parameters specific to the component under consideration. These parameters must be adjusted to experimental data (saturated vapor pressure, heat capacity, etc.). Among the most common specific functions, we find, for example, the Twu function:

$$\alpha_{Twu}(T_r) = T_r^{N(M-1)} e^{L\left(1-T_r^{MN}\right)}$$

[2.20]

The three adjustable parameters are denoted as L, M and N. Another example is the Mathias–Copeman function:

$$\alpha_{Mathias-Copeman}(T_r) = \left[1 + c_1\left(1 - \sqrt{T_r}\right) + c_2\left(1 - \sqrt{T_r}\right)^2 + c_3\left(1 - \sqrt{T_r}\right)^3\right]^2$$

[2.21]

This function also includes three adjustable parameters denoted as c_1, c_2 and c_3.

– Capacity of the cubic equations of state to correlate non-volumetric properties of pure components:

Based on recent studies that we have carried out on around a thousand pure substances, we have been able to show that cubic equations of state combined with the generalized Soave function allow the saturated vapor pressures to be reproduced on average to the nearest 2%, vaporization enthalpies to the nearest 3% and heat capacities to the nearest 7–8%.

Using the specific Twu function, the accuracies can reach on average 1% on saturated vapor pressures, 2% on vaporization enthalpies and 2% on heat capacities.

– Capacity of cubic equations of state to reproduce the volumetric properties of pure substances:

This is a known weakness of the classic cubic equations of state: they do not allow the liquid densities of pure substances and mixtures to be reproduced accurately. We should note that, on average, the SRK equation induces an error of around 17%, whereas for PR, the error is only 7%.

2.3.3. Introduction of volume translation in cubic equations of state with the aim of improving estimation of liquid densities

To partially resolve the problem of incorrect restitution of the liquid densities by classic cubic equations of state (see Figure 2.3), a method consists of translating all the calculated volumes of a constant quantity known as volume correction, which is generally denoted as c (for *correction*) or t (for *translation*).

If v_{ori} and b_{ori} designate the molar volume and the covolume calculated by the original cubic equation of state, then:

$$\begin{cases} \underbrace{v_{corrected}}_{\substack{\text{to be compared} \\ \text{to experimental data}}} = v_{ori} - c \\[2em] b_{corrected} = b_{ori} - c \end{cases} \qquad [2.22]$$

As an example, let us apply this volume correction to the SRK equation.

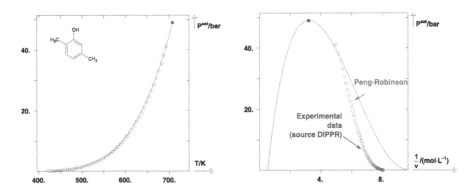

Figure 2.3. *Prediction of the volumetric properties using a cubic equation of state. For a color version of the figure, see www.iste.co.uk/jaubert/thermodynamic.zip*

The original equation:

$$P(T, v_{ori}) = \frac{RT}{v_{ori} - b_{ori}} - \frac{a(T)}{v_{ori}(v_{ori} + b_{ori})}$$ [2.23]

becomes:

$$P(T, v_{corrected}) = \frac{RT}{v_{corrected} - b_{corrected}} - \frac{a(T)}{(v_{corrected} + c)(v_{corrected} + b_{corrected} + 2c)}$$

[2.24]

This operation is graphically illustrated in Figure 2.4.

The volume correction parameter of an equation of state for pure substances should preferably be chosen temperature-independent in order to avoid isotherm crossing in the pressure–molar volume plane, which is not observed experimentally, as well as to avoid predicting negative heat capacities at high temperatures, in violation of the thermal stability criterion. On the other hand, this parameter is not universal: it is specific to the component in question. Two main methods can be used to estimate it:

(1) Universal correlations have been developed. Their use requires the knowledge of the Rackett compressibility factor z_{RA} (see Chapter 1). As a

reminder, these factors are used to estimate the liquid molar volumes through the Rackett correlation. For example, Péneloux proposes:

$$\begin{cases} \text{PR:} \quad c = \dfrac{RT_{c,\exp}}{P_{c,\exp}}\left(0.1154 - 0.4406 \cdot z_{RA}\right) \\[4mm] \text{SRK:} \quad c = \dfrac{RT_{c,\exp}}{P_{c,\exp}}\left(0.1156 - 0.4077 \cdot z_{RA}\right) \end{cases} \qquad [2.25]$$

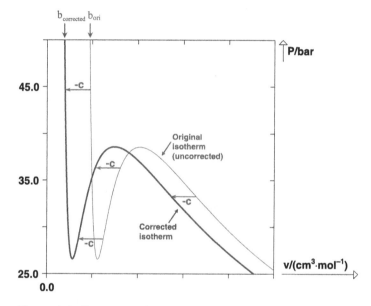

Figure 2.4. *Illustration of the concept of volume translation: all the molar volumes are shifted by a constant value* c

(2) If an experimental value of liquid molar volume at a given temperature $T_{réf}$ is known, we can estimate c using:

$$c = v_{L,ori}^{sat}(T_{ref}) - v_{L,\exp}^{sat}(T_{ref}) \qquad [2.26]$$

where $v_{L,ori}^{sat}$ designates the liquid molar volume at saturation that is predicted by the untranslated equation of state.

For example, in Figure 2.5, we show the effect of a volume translation on the restitution of liquid densities.

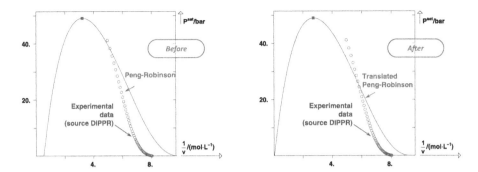

Figure 2.5. *Effect of a volume translation on the restitution of liquid densities of pure ethane at saturation*

This technique presents two advantages. On the one hand, **it leads to phenomenal improvements in estimates of liquid densities**: for a thousand subcritical compounds, the average error on the estimation of liquid molar volumes is approximately 2.5% for the PR equation (this is three times less than with the untranslated equation) and 4.0% for the SRK equation (i.e. four times less than with the untranslated equation). In addition, **the volume translation operation does not affect the calculated values of the vapor–liquid equilibrium properties** (saturated vapor pressures, vaporization enthalpies, heat capacities at saturation, etc.).

As a conclusion to the paragraphs devoted to the application of cubic equations of state to pure substances, we can assert that these models:

– are applicable to a large number of chemical families: hydrocarbons, amines, esters, permanent gases, refrigerants, etc. Although not as good as those obtained for other types of compounds, the results obtained with highly polar compounds (and in particular those associating through hydrogen bonds) such as alcohols remain acceptable. By contrast, the predictions can deteriorate very quickly in the case of acids (due to the phenomenon of dimerization);

– lead to very good estimates of the properties $P^{sat}(T)$, $\Delta_{vap}H(T)$ and $c_{P,L}^{sat}(T)$ on condition that the function $\alpha(T)$ of the attractive term is correctly parameterized;

– give unsatisfactory predictions of liquid densities that can be partially overcome by using a volume translation parameter c.

NOTE.– Some cubic equations of state use a covolume b that depends on the temperature. In the same way as the volume translation parameter, this parameter must remain independent of the temperature in order to avoid theoretical problems such as prediction of negative heat capacities and isotherm crossings in the pressure–molar volume plane.

2.3.4. Equations of state relying on the statistical associating fluid theory (SAFT)

The first "SAFT" equation of state was proposed in 1989 by Chapman et al.[3]. It is based on a statistical-thermodynamics theory known as TPT (thermodynamic perturbation theory) developed by Wertheim. This equation and all the variants that followed express the Helmholtz energy a of a fluid. As in molecular simulation, a non-associating fluid is seen as an assembly (or a chain) of m fictive segments (or monomers), each segment being characterized by its diameter (σ) and a dispersion energy (ε). In simple cases (linear alkanes, for example), a segment can be interpreted as an atom or a group of atoms of the real molecule, but in most cases, a segment is the constitutive unit of an equivalent fictional molecule, composed of m identical monomers, with the same properties as the real molecule. The expression of molar Helmholtz energy a of the real fluid summarizes the energies required for formation of the chain:

$$a = a^{\bullet} + \underbrace{a^{res,HS} + a^{res,Disp\,seg}}_{a^{res,seg}} + a^{res,Chain\,of\,seg} + a^{res,Assoc} \qquad [2.27]$$

where:

$a^{\bullet}$ designates the Helmholtz energy of a perfect gas with the same temperature and the same molar volume as the real fluid;

$a^{res,HS}$ designates the residual Helmholtz energy to be provided to the perfect gas to obtain the Helmholtz energy of a fluid composed of hard spheres (the hard spheres are non-deformable hypothetical particles, subject to repulsive forces when they come into contact with each other but not

---

3 W.G. Chapman, K.E. Gubbins, G. Jackson, M. Radosz, *Fluid Phase Equilibria*, 52, pp. 31–38, 1989.

subject to attractive forces). Thus, $a^{\bullet} + a^{res,HS}$ represents the Helmholtz energy of a fluid composed of hard spheres.

The term $a^{res,Disp\ seg}$ introduces dispersive (attractive) forces. Based on the same principle as before, the quantity $a^{\bullet} + a^{res,HS} + a^{res,Disp\ seg}$ represents the Helmholtz energy of a fluid composed of monomers (or segments) that are subject to repulsive and attractive forces. We note that the sum $a^{res,HS} + a^{res,Disp\ seg}$ is related to the residual Helmholtz energy of a segment ($a^{res,seg}$).

Adding the term $a^{res,Chain\ of\ seg}$, we summarize the formation of a chain of monomers (formation of covalent bonds between the segments).

Finally, the term $a^{res,Assoc}$ expresses the presence of association by hydrogen bonding that is the electrostatic interaction between a partial positive charge (δ^+) located on a labile hydrogen atom and a partial negative charge (δ^-) located on an atom that possesses a lone pair of electrons.

The mathematical expression of the corresponding pressure-explicit equation of state is obtained by simply deriving equation [2.27]:

$$P(T,v) = -\left(\frac{\partial a}{\partial T}\right)_v \qquad [2.28]$$

Many SAFT-type equations of state exist in the literature. Among these, we can cite SAFT-VR, SOFT-SAFT and PC-SAFT as probably the main examples.

In general, these equations of state require three input parameters to characterize a pure non-associating fluid. If the fluid is associating, two additional parameters per association site must be known, as summarized in Figure 2.6. We should note that in contrast to the cubic equations in which the values of universal parameters Ω_a and Ω_b are set so that the equation of state exactly reproduces the critical coordinates $T_{c,exp}$ and $P_{c,exp}$, SAFT equations are generally parameterized differently. The three parameters that are not attached to the association term (m, σ and ε in Figure 2.6) are often determined in order to minimize the deviations between calculated and experimental vapor pressures and liquid densities.

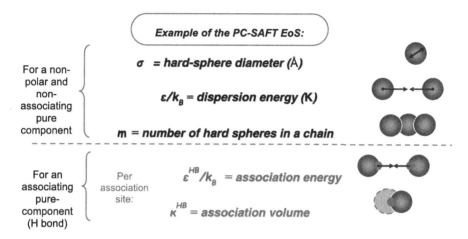

Figure 2.6. *Presentation of the input parameters of the equation PC-SAFT for pure substances. EoS means "equation of state". For a color version of the figure, see www.iste.co.uk/jaubert/thermodynamic.zip*

Reproduction of critical coordinates is never taken into account. These different ways of parameterizing models induce different behaviors that we are now going to discuss. We begin by setting out the main strengths and weaknesses of the SAFT equations:

Strengths of SAFT equations of state: the liquid densities are better correlated than with a cubic equation of state. The properties of associating and highly polar fluids are better correlated than with a cubic equation of state. We also know by experience that these models are suitable for describing polymers.

Weaknesses of SAFT equations of state: there are no (or very few) general and systematic methods to estimate the input parameters of these equations of state. In addition, as a result of the way in which these models are parameterized, the critical pressures are not restituted well, in particular for heavy molecules (on the contrary to cubic equations).

Figure 2.7 gives an illustration of the performances of the PC-SAFT equation of state that is used to model the saturated vapor pressure and the liquid density of ethylbenzene at saturation.

As a conclusion to this paragraph and to those concerning the cubic equations of state, we will recall that:

– the **cubic equations** are parameterized to reproduce exactly $T_{c,\text{exp}}$ and $P_{c,\text{exp}}$ but not to reproduce the liquid densities;

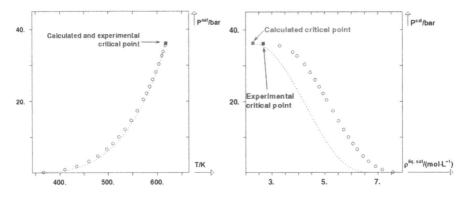

Figure 2.7. *Modeling of the properties of vapor–liquid equilibrium of ethylbenzene by the PC-SAFT equation*

– the **SAFT equations** are generally parameterized to reproduce the experimental data of saturated vapor pressure and liquid densities. They are not constrained to reproduce $T_{c,\text{exp}}$ and $P_{c,\text{exp}}$. Of course, if the SAFT equations were constrained to reproduce $T_{c,\text{exp}}$ and $P_{c,\text{exp}}$, prediction of the liquid densities would be significantly inferior.

2.3.5. *Equations of state specific to particular pure substances*

It appears that equations of state presented above (virial, cubic, SAFT) are universal in the sense that these models can be applied to an infinity of molecules: by modifying model input parameters, we switch from one molecule to another. These parameters are generally experimental properties that are easily accessible or can be estimated by simple optimization procedures. It is even sometimes possible to predict the values of input parameters (from knowledge of another property, from the molecular structure, etc.).

In the open literature, we also find equations of state that are specific to certain components and which were developed in order to model the thermodynamic properties of these components to a high degree of precision. They are not universal and not adaptable to another compound; they generally contain a very large number of component parameters.

For example, the NIST (National Institute of Standards and Technology) developed a software program known as REFPROP (Reference Fluid Thermodynamic and Transport Properties Database), which contains specific equations of state for 121 pure substances. In this category of models, we can cite, for example, the Benedict–Webb–Rubin (BWR) equations of state, Bender, Span–Wagner, GERG-2008, etc.

Figure 2.8, which compares the aptitude of several equations of state to represent the speed of sound in pure CO_2, illustrates the accuracy that can be reached with this type of model. Here, the Span–Wagner (SW) equation enables the best reproduction of the experimental data.

We should remember that the great strength of these models is their accuracy, whereas their weakness lies in their poor flexibility: each pure substance is represented by an equation of state that involves a large number of fitted parameters (generally several dozen). Processing of the mixtures also requires knowledge of additional interaction parameters. The use of these models is therefore limited to pure substances and mixtures that have previously been studied by their developers, which remains an obstacle to their use.

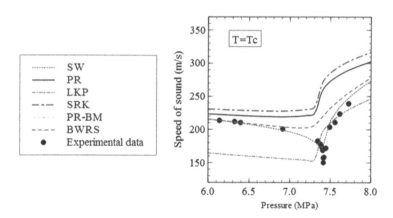

Figure 2.8. *Modeling of the celerity of sound in pure CO_2 using different approaches. Non-specific equations of state: PR = Peng–Robinson, LKP = Lee–Kesler–Plöcker, SRK = Soave–Redlich–Kwong, PR-BM = Peng–Robinson–Boston–Mathias. Equations of state that are specific to CO_2: SW = Span–Wagner, BWRS = Benedict–Webb–Rubin–Starling. For a color version of the figure, see www.iste.co.uk/jaubert/thermodynamic.zip*

By way of illustration, the GERG-2008 model was developed to represent natural gas mixtures. This model applies to any fluid containing one or several of the molecules listed here: methane, nitrogen, CO_2, ethane, propane, n-butane, isobutene, n-pentane, isopentane, n-hexane, n-heptane, n-octane, n-nonane, n-decane, hydrogen, oxygen, CO, water, H_2S, helium, argon.

2.4. Practical use of volumetric equations of state

2.4.1. *Calculation of the state properties*

The equation of state allows calculation of the correction that must be made to the property of the perfect gas in order to obtain the property of the real fluid. We will therefore write:

$$\underbrace{x^*}_{\substack{\text{property of} \\ \text{real fluid}}} = \underbrace{x^\bullet}_{\substack{\text{property of} \\ \text{perfect gas}}} + \underbrace{x^{correction}}_{\substack{\text{given by the} \\ \text{equation of state}}} \qquad [2.29]$$

The symbol * is used to indicate that the property belongs to the pure real fluid, whereas the symbol $^\bullet$ designates a perfect-gas property. The properties x that can be calculated using an equation of state are, for example, enthalpy, entropy, internal energy, Gibbs energy, Helmholtz energy, heat capacities, speed of sound, etc.

2.4.2. *Vapor–liquid equilibrium*

When the equation of state is applicable to both the liquid and vapor phases, it allows the **conditions of equilibrium between fluid phases to be resolved** and to calculate the associated state properties. The saturated vapor pressure, boiling point temperature, vaporization property, and so on of a pure substance can thus be estimated. Phase diagrams such as those presented in Figure 2.9 can thus also be calculated.

2.5. Calculation of state properties using a volumetric equation of state

The state properties are calculated from equation [2.29]. However, this equation needs to be clarified. Effectively, if the real fluid is in a $\left(P^*, T^*, v^* \right)$ state, the perfect gas cannot be in the same state. This is because a real fluid and a perfect gas do not obey the same equation of state. It is noteworthy that **a real fluid and a perfect gas will have at most two state variables in common among pressure, molar volume and temperature (the third variable – which will be different – is set by the equation of state of each of the two fluids).** Consequently, to any real fluid state, it is possible to associate three perfect gas states as indicated in Figure 2.10.

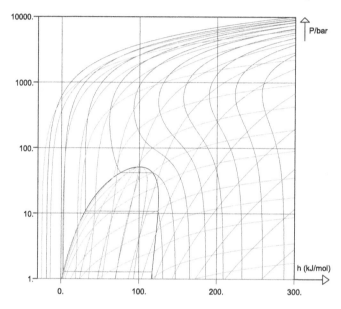

Figure 2.9. *Phase diagram for CO_2 represented in the pressure–molar enthalpy plane calculated by the PR equation of state and to which isovalue lines have been added (isenthalpes, isobars, isentropes, isochores, iso-quality). For a color version of the figure, see www.iste.co.uk/jaubert/thermodynamic.zip*

For convenience (simplicity of calculations), the choice of the *perfect gas state* is dictated by the mathematical form of the equation of state.

If the equation of state is volume-explicit, i.e. has the form $v^*(T,P) = \ldots$ or in other words, its working variables are T and P, the perfect gas no. 1 will be chosen (see Figure 2.10). The correction to the perfect gas is known as the **residual-TP property**. Thus:

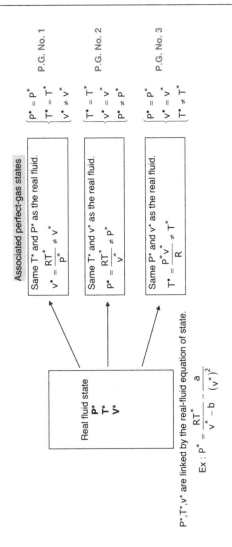

Figure 2.10. *Presentation of the three "perfect gas" (P.G.) states associated with a real fluid state*

$$\underbrace{x^*(T^*,P^*)}_{\substack{\text{property of}\\\text{the real fluid}\\(v^*\text{ given by the}\\\text{equation of state}\\\text{of the real fluid})}} = \underbrace{x^\bullet(T^*,P^*)}_{\substack{\text{property of}\\\text{the perfect gas}\\v^\bullet=(RT^*/P^*)\neq v^*}} + \underbrace{x^{res-TP}(T^*,P^*)}_{\substack{\text{residual-T,P property}\\\text{given by the volume-explicit}\\\text{equation of state of the real fluid}}} \qquad [2.30]$$

If the equation of state is pressure-explicit, i.e. has the form $P^*(T,v) = \ldots$ or in other words, its working variables are T and v, the perfect gas no. 2 will be chosen (see Figure 2.10). The correction to the perfect gas is known as **residual-TV property**. Thus:

$$\underbrace{x^*(T^*,v^*)}_{\substack{\text{property of} \\ \text{the real fluid} \\ (P^* \text{ given by the} \\ \text{equation of state} \\ \text{of the real fluid})}} = \underbrace{x^\bullet(T^*,v^*)}_{\substack{\text{property of} \\ \text{the perfect gas} \\ P^\bullet = (RT^*/v^*) \neq P^*}} + \underbrace{x^{res-TV}(T^*,v^*)}_{\substack{\text{residual-T,v property} \\ \text{given by the pressure-explicit} \\ \text{equation of state of the real fluid}}}$$

[2.31]

2.5.1. *Calculation of residual-TP properties*

When expressed in variables (T,P), "g" (the Gibbs energy) is a so-called *characteristic function*. This means that knowledge of the function $g^*(T,P)$ allows all other state functions to be calculated in the same set of variables, as shown in equation [2.32].

$$\begin{cases} s(T,P) = -\left[\dfrac{\partial g(T,P)}{\partial T}\right]_P \\[2mm] v(T,P) = \left[\dfrac{\partial g(T,P)}{\partial P}\right]_T \\[2mm] h(T,P) = g(T,P) + T \cdot s(T,P) \\[2mm] c_P(T,P) = \left[\dfrac{\partial h(T,P)}{\partial T}\right]_P \\[2mm] u(T,P) = h(T,P) - P \cdot v(T,P) \\[2mm] a(T,P) = g(T,P) - P \cdot v(T,P) \\[2mm] c_v(T,P) = c_P(T,P) + T\left[\left(\dfrac{\partial v(T,P)}{\partial T}\right)_P\right]^2 \Big/ \left(\dfrac{\partial v(T,P)}{\partial P}\right)_T \end{cases}$$

[2.32]

Consequently, it is possible to state that if we can calculate $g^{res-TP}(T,P)$, the other residual-TP properties will be calculated according to the procedure described in equation [2.32]. By definition:

$$g^{res-TP}(T,P) = g^*(T,P) - g^{\bullet}(T,P)$$ [2.33]

When the pressure approaches zero, the real fluid approaches the perfect gas state and:

$$g^*(T,P=0) = g^{\bullet}(T,P=0)$$ [2.34]

Thus:

$$
\begin{aligned}
g^{res-TP}(T,P) &= \left[g^*(T,P) - g^*(T,P=0) \right] - \left[g^{\bullet}(T,P) - g^{\bullet}(T,P=0) \right] \\
&= \int_0^P v^*(T,P) \cdot dP - \int_0^P v^{\bullet}(T,P) \cdot dP \\
&= \int_0^P \left[\underbrace{v^*(T,P)}_{equation\ of\ state} - \frac{RT}{P} \right] \cdot dP \\
&= \int_0^P v^{res-TP}(T,P) \cdot dP
\end{aligned}
$$ [2.35]

DEFINITION.– The fugacity coefficient φ of the real fluid is defined by $\ln \varphi = g^{res-TP}/(RT)$.

We will now illustrate calculation of the residual-TP properties, taking as an example the truncated virial equation of state. According to equation [2.9], we have $v^*(T,P) = \dfrac{RT}{P} + B(T)$ and thus, that

$v^{res-TP}(T,P) = v^*(T,P) - \dfrac{RT}{P} = B(T)$ and all other residual-TP properties are given by equation [2.36].

$$\left\{\begin{array}{l}
g^{res-TP}(T,P) = \int_0^P v^{res-TP}(T,P) \cdot dP = B(T) \cdot P \\[2mm]
s^{res-TP}(T,P) = -\left[\dfrac{\partial g^{res-TP}(T,P)}{\partial T}\right]_P = -B'(T) \cdot P \quad \text{(with } B' = dB/dT) \\[2mm]
h^{res-TP}(T,P) = g^{res-TP}(T,P) + T \cdot s^{res-TP}(T,P) = P(B - TB') \\[2mm]
c_P^{res-TP}(T,P) = \left[\dfrac{\partial h^{res-TP}(T,P)}{\partial T}\right]_P = -TPB'' \quad \text{(with } B'' = d^2B/dT^2) \\[2mm]
u^{res-TP}(T,P) = h^{res-TP}(T,P) - P \cdot v^{res-TP}(T,P) = -TPB' \\[2mm]
a^{res-TP}(T,P) = g^{res-TP}(T,P) - P \cdot v^{res-TP}(T,P) = 0 \\[2mm]
c_v^{res-TP}(T,P) = c_P^{res-TP}(T,P) + R + T\left[\left(\dfrac{\partial v^*(T,P)}{\partial T}\right)_P\right]^2 \bigg/ \left(\dfrac{\partial v^*(T,P)}{\partial P}\right)_T \\[2mm]
\qquad\qquad = -TPB'' + R - \dfrac{P^2 T}{R}\left(\dfrac{R}{P} + B'\right)^2
\end{array}\right. \qquad [2.36]$$

2.5.2. Calculation of residual-TV properties

When expressed in variables (T,v), "a" (the Helmholtz energy) is a so-called characteristic function. This means that knowledge of $a^*(T,v)$ allows all other state functions to be calculated in the same set of variables as shown in equation [2.37].

$$\left\{\begin{array}{l}
s(T,v) = -\left[\dfrac{\partial a(T,v)}{\partial T}\right]_v \\[2mm]
P(T,v) = -\left[\dfrac{\partial a(T,v)}{\partial v}\right] \\[2mm]
u(T,v) = a(T,v) + T \cdot s(T,v) \\[2mm]
c_v(T,v) = \left[\dfrac{\partial u(T,v)}{\partial T}\right]_v \\[2mm]
h(T,v) = u(T,v) + v \cdot P(T,v) \\[2mm]
g(T,v) = a(T,v) + v \cdot P(T,v) \\[2mm]
c_P(T,v) = c_v(T,v) - T\left[\left(\dfrac{\partial P(T,v)}{\partial T}\right)_v\right]^2 \bigg/ \left(\dfrac{\partial P(T,v)}{\partial v}\right)_T
\end{array}\right. \qquad [2.37]$$

Consequently, it is possible to state that if we can calculate $a^{res-TV}(T,v)$, the other residual-TV properties will be calculated according to the procedure described in equation [2.37]. By definition:

$$a^{res-TV}(T,v) = a^*(T,v) - a^\bullet(T,v) \qquad [2.38]$$

When the molar volume approaches $+\infty$, the real fluid approaches the perfect gas state and:

$$a^*(T,v=+\infty) = a^\bullet(T,v=+\infty) \qquad [2.39]$$

Thus:

$$
\begin{aligned}
a^{res-TV}(T,v) &= \left[a^*(T,v) - a^*(T,v=+\infty)\right] - \left[a^\bullet(T,v) - a^\bullet(T,v=+\infty)\right] \\
&= -\int_{+\infty}^{v} P^*(T,v)\cdot dv + \int_{+\infty}^{v} P^\bullet(T,v)\cdot dv \\
&= -\int_{+\infty}^{v} \left[\underbrace{P^*(T,v)}_{equation\ of\ state} - \frac{RT}{v}\right]\cdot dv \\
&= -\int_{+\infty}^{v} -P^{res-TV}(T,v)\cdot dv
\end{aligned}
$$

$$[2.40]$$

We will now illustrate calculation of the residual-TV properties, taking as an example the cubic equations of state (in their generalized form given by equation [2.11]). They are written as: $P^*(T,v) = \dfrac{RT}{v-b} - \dfrac{a(T)}{(v-b\cdot r_1)(v-b\cdot r_2)}$.
Two cases must be distinguished: the case $r_1 \neq r_2$ (this is the case of SRK or PR models; see equation [2.11]) and the case $r_1 = r_2$ (this is the case of the Van der Waals model; see equation [2.11]).

In the case $r_1 \neq r_2$, we have immediately: $P^{res-TV}(T,v) = \dfrac{RT}{v-b}$

$-\dfrac{a(T)}{(v-b\cdot r_1)(v-b\cdot r_2)} - \dfrac{RT}{v}$ in such a way that all the other residual-TV properties are given by equation [2.41].

$$a^{res-TV}(T,v) = -\int_{+\infty}^{v} P^{res-TV}(T,v) \cdot dv$$

$$= RT \ln\left(\frac{v}{v-b}\right) + \frac{a(T)}{b \cdot (r_1 - r_2)} \ln\left(\frac{v-b \cdot r_1}{v-b \cdot r_2}\right)$$

$$s^{res-TV}(T,v) = -\left[\frac{\partial a^{res-TV}(T,v)}{\partial T}\right]_v$$

$$= R \cdot \ln\left(\frac{v-b}{v}\right) - \frac{1}{b \cdot (r_1 - r_2)} \cdot \frac{da}{dT} \cdot \ln\left(\frac{v-b \cdot r_1}{v-b \cdot r_2}\right)$$

$$u^{res-TV}(T,v) = a^{res-TV}(T,v) + T \cdot s^{res-TV}(T,v)$$

$$= \frac{1}{b \cdot (r_1 - r_2)} \cdot \left(a - T \cdot \frac{da}{dT}\right) \cdot \ln\left(\frac{v-b \cdot r_1}{v-b \cdot r_2}\right)$$

$$c_v^{res-TV}(T,v) = \left[\frac{\partial u^{res-TV}(T,v)}{\partial T}\right]_v = \frac{-T}{b \cdot (r_1 - r_2)} \cdot \frac{d^2 a}{dT^2} \cdot \ln\left(\frac{v-b \cdot r_1}{v-b \cdot r_2}\right)$$

$$h^{res-TV}(T,v) = u^{res-TV}(T,v) + v \cdot P^{res-TV}(T,v)$$

$$= \frac{1}{b \cdot (r_1 - r_2)} \cdot \left(a - T \cdot \frac{da}{dT}\right) \cdot \ln\left(\frac{v-b \cdot r_1}{v-b \cdot r_2}\right)$$

$$+ \frac{R \cdot T \cdot b}{v-b} - \frac{a(T) \cdot v}{(v-b \cdot r_1)(v-b \cdot r_2)}$$

$$g^{res-TV}(T,v) = a^{res-TV}(T,v) + v \cdot P^{res-TV}(T,v)$$

$$= RT \ln\left(\frac{v}{v-b}\right) + \frac{a(T)}{b \cdot (r_1 - r_2)} \ln\left(\frac{v-b \cdot r_1}{v-b \cdot r_2}\right)$$

$$+ \frac{R \cdot T \cdot b}{v-b} - \frac{a(T) \cdot v}{(v-b \cdot r_1)(v-b \cdot r_2)}$$

$$c_P^{res-TV}(T,v) = c_v^{res-TV}(T,v) - R - T\left[\left(\frac{\partial P^*(T,v)}{\partial T}\right)_v\right]^2 \bigg/ \left(\frac{\partial P^*(T,v)}{\partial v}\right)_T$$

with:
$$\left(\frac{\partial P^*(T,v)}{\partial T}\right)_v = \frac{R}{v-b} - \frac{da}{dT} \cdot \frac{1}{(v-b \cdot r_1)(v-b \cdot r_2)}$$

and:
$$\frac{-1}{\left(\dfrac{\partial P^*(T,v)}{\partial v}\right)_T} = \frac{1}{\dfrac{RT}{(v-b)^2} - \dfrac{a(T) \cdot [2v - (r_1 + r_2) \cdot b]}{(v-b \cdot r_1)^2 (v-b \cdot r_2)^2}}$$

$$[2.41]$$

In the case $r_1 = r_2$ (hereafter simply denoted as r), $P^{res-TV}(T,v) =$ $\dfrac{RT}{v-b} - \dfrac{a(T)}{(v-b\cdot r)^2} - \dfrac{RT}{v}$ in such a way that all other residual-TV properties are given by equation [2.42].

$$
\left|
\begin{array}{l}
a^{res-TV}(T,v) = -\displaystyle\int_{+\infty}^{v} P^{res-TV}(T,v) \cdot dv = RT \ln\left(\dfrac{v}{v-b}\right) - \dfrac{a(T)}{v-br} \\[12pt]
s^{res-TV}(T,v) = -\left[\dfrac{\partial a^{res-TV}(T,v)}{\partial T}\right]_v = R \cdot \ln\left(\dfrac{v-b}{v}\right) + \dfrac{1}{v-br} \cdot \dfrac{da}{dT} \\[12pt]
u^{res-TV}(T,v) = a^{res-TV}(T,v) + T \cdot s^{res-TV}(T,v) = \dfrac{-1}{v-br} \cdot \left(a - T \cdot \dfrac{da}{dT}\right) \\[12pt]
c_v^{res-TV}(T,v) = \left[\dfrac{\partial u^{res-TV}(T,v)}{\partial T}\right]_v = \dfrac{T}{(v-br)} \cdot \dfrac{d^2 a}{dT^2} \\[12pt]
h^{res-TV}(T,v) = u^{res-TV}(T,v) + v \cdot P^{res-TV}(T,v) \\[12pt]
\qquad = -\dfrac{1}{(v-br)} \cdot \left(a - T \cdot \dfrac{da}{dT}\right) + \dfrac{R \cdot T \cdot b}{v-b} - \dfrac{a(T) \cdot v}{(v-br)^2} \\[12pt]
g^{res-TV}(T,v) = a^{res-TV}(T,v) + v \cdot P^{res-TV}(T,v) \\[12pt]
\qquad = RT \ln\left(\dfrac{v}{v-b}\right) - \dfrac{a(T)}{v-br} + \dfrac{R \cdot T \cdot b}{v-b} - \dfrac{a(T) \cdot v}{(v-br)^2} \\[12pt]
c_P^{res-TV}(T,v) = c_v^{res-TV}(T,v) - R - T\left[\left(\dfrac{\partial P^*(T,v)}{\partial T}\right)_v\right]^2 \bigg/ \left(\dfrac{\partial P^*(T,v)}{\partial v}\right)_T \\[12pt]
\text{with: } \left(\dfrac{\partial P^*(T,v)}{\partial T}\right)_v = \dfrac{R}{v-b} - \dfrac{da}{dT} \cdot \dfrac{1}{(v-br)^2} \\[12pt]
\text{and: } \dfrac{-1}{\left(\dfrac{\partial P^*(T,v)}{\partial v}\right)_T} = \dfrac{1}{\dfrac{RT}{(v-b)^2} - \dfrac{2a(T)}{(v-br)^3}}
\end{array}
\right.
$$

$$[2.42]$$

2.5.3. Calculation of state-property changes

In this section, we demonstrate that a state-property change can be calculated for a real fluid if we know its equation of state and if we have an expression for the **molar heat capacity of the perfect gas** (denoted as $c_P^\bullet(T)$). In light of the previous sections, two cases must therefore be distinguished.

On the one hand, if the equation of state is volume-explicit:

$$\Delta x^*_{\substack{T_1 \to T_2 \\ P_1 \to P_2 \\ (v_1 \to v_2)}} = \Delta x^\bullet_{\substack{T_1 \to T_2 \\ P_1 \to P_2 \\ (v_1^\bullet \to v_2^\bullet)}} + \underbrace{\Delta x^{res-TP}_{\substack{T_1 \to T_2 \\ P_1 \to P_2 \\ P_1 \to P_2}}}_{\substack{equation \\ of\ state}} \qquad [2.43]$$

with:

$$\left\{ \begin{array}{l}
\Delta h^\bullet_{\substack{T_1 \to T_2 \\ P_1 \to P_2}} = \int_{T_1}^{T_2} c_P^\bullet(T) \cdot dT \\[3mm]
\Delta u^\bullet_{\substack{T_1 \to T_2 \\ P_1 \to P_2}} = \int_{T_1}^{T_2} \left[c_P^\bullet(T) - R \right] \cdot dT \\[3mm]
\Delta s^\bullet_{\substack{T_1 \to T_2 \\ P_1 \to P_2}} = \int_{T_1}^{T_2} \frac{c_P^\bullet(T)}{T} \cdot dT - R \ln\left(\frac{P_2}{P_1} \right) \\[3mm]
\Delta c_P^\bullet_{\substack{T_1 \to T_2 \\ P_1 \to P_2}} = \Delta c_v^\bullet_{\substack{T_1 \to T_2 \\ P_1 \to P_2}} = c_P^\bullet(T_2) - c_P^\bullet(T_1)
\end{array} \right. \qquad [2.44]$$

In particular, we should note that the molar enthalpy, the internal molar energy and the molar heat capacities of a perfect gas only depend on the temperature (in light of Joule's laws). This is the meaning of the indications $P_1 \to P_2$ in equation [2.44].

On the other hand, if the equation of state is pressure-explicit:

$$\Delta x^*_{\substack{T_1 \to T_2 \\ v_1 \to v_2 \\ (P_1 \to P_2)}} = \Delta x^\bullet_{\substack{T_1 \to T_2 \\ v_1 \to v_2 \\ (P_1^\bullet \to P_2^\bullet)}} + \underbrace{\Delta x^{res-TV}_{\substack{T_1 \to T_2 \\ v_1 \to v_2}}}_{\substack{equation \\ of\ state}} \qquad [2.45]$$

with:

$$
\begin{cases}
\underset{\substack{T_1 \to T_2 \\ v_1 \to v_2}}{\Delta h^\bullet} = \int_{T_1}^{T_2} c_p^\bullet(T) \cdot dT \\[2em]
\underset{\substack{T_1 \to T_2 \\ v_1 \to v_2}}{\Delta u^\bullet} = \int_{T_1}^{T_2} \left[c_p^\bullet(T) - R \right] \cdot dT \\[2em]
\underset{\substack{T_1 \to T_2 \\ v_1 \to v_2}}{\Delta s^\bullet} = \int_{T_1}^{T_2} \frac{c_p^\bullet(T)}{T} \cdot dT + R \ln \left(\frac{T_1 \cdot v_2}{T_2 \cdot v_1} \right) \\[2em]
\underset{\substack{T_1 \to T_2 \\ v_1 \to v_2}}{\Delta c_p^\bullet} = \underset{\substack{T_1 \to T_2 \\ v_1 \to v_2}}{\Delta c_v^\bullet} = c_p^\bullet(T_2) - c_p^\bullet(T_1)
\end{cases}
$$

[2.46]

As mentioned previously, the indications $v_1 \to v_2$ in equation [2.46] provide a reminder that the molar enthalpy, the molar internal energy and the molar heat capacities of a perfect gas only depend on temperature.

2.6. Vapor–liquid equilibrium calculation using a pressure-explicit equation of state: illustration with cubic equations of state

2.6.1. Shapes of isotherms produced by cubic equations of state in the pressure–molar volume plane: link between the form of isotherms and the resolution of cubic equations of state

The shapes of subcritical, critical and supercritical isotherms of pure substances in the (P, v) plane produced from experimental measurements are shown in Figure 1.5. Those produced by the cubic equations of state are illustrated in Figure 2.11. It is therefore observed that the subcritical (or *hypocritical*) isotherms produced by the cubic equations all have a local maximum and a local minimum. Graph curves of this type are sometimes named ***Van der Waals isotherms***. These singularities have consequences on the resolution of equations of state at fixed temperature and pressure (i.e. when seeking the molar volumes associated with these conditions).

As explained in section 2.3.2, the term *"cubic"* means that for a given temperature and pressure, the molar volume is the solution to a polynomial equation of degree three. We will now prove this characteristic from the general form of cubic equations of state.

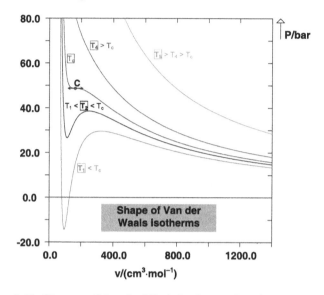

Figure 2.11. *Shapes of Van der Waals isotherms for ethane produced by the Peng–Robinson equation of state, in the (P, v) plane*

$$P = \frac{RT}{v-b} - \frac{a(T)}{(v-b \cdot r_1)(v-b \cdot r_2)}$$

$$\Leftrightarrow \quad v^3 - \left[b(r_1 + r_2 + 1) + \frac{RT}{P} \right] v^2$$

$$+ \left[b^2(r_1 r_2 + r_1 + r_2) + \frac{RTb}{P}(r_1 + r_2) + \frac{a}{P} \right] v$$

$$- b \left(r_1 r_2 b^2 + \frac{r_1 r_2 bRT}{P} + \frac{a}{P} \right) = 0$$

[2.47]

As a reminder, the ranges of validity of the cubic equations of state are $\begin{cases} T > 0 \\ v > b \end{cases}$ (see equation [2.11]). The number of roots (i.e. solutions) of the polynomial equation [2.47] in the $v > b$ range depends on the values for the fixed temperature and pressure and can be discussed graphically using Figure 2.12. Here, we observe that at fixed temperature and pressure, a cubic equation of state can have one or three real roots (the case of two real roots corresponding to a situation where the pressure would be set at the local minimum or the local maximum value is numerically improbable). When the equation of state has three roots:

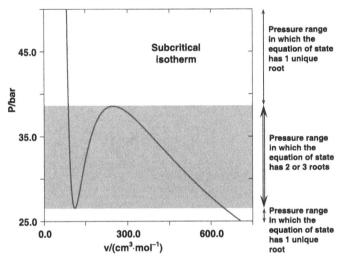

Figure 2.12. *Graphical discussion of the number of roots of a cubic equation of state for a pure substance at a fixed subcritical temperature and fixed pressure*

– the smallest molar volume is declared *liquid*;

– the largest molar volume is declared *gas*;

– the intermediate molar volume is said to be *unstable*.

By extending this, we can define three branches on a Van der Waals isotherm, one associated with liquid states, the second associated with gaseous states and the last with states known as *unstable*. These branches are represented in Figure 2.13.

2.6.2. *Stable, metastable, unstable roots*

The criterion for mechanical stability (imposed by the second law of thermodynamics) states that $(\partial P / \partial v)_T \leq 0$. A state that violates this criterion is declared unstable. All the states on the unstable branch of the isotherm in Figure 2.13 such that $(\partial P / \partial v)_T > 0$ have therefore been declared unstable.

In the previous chapter, the saturated vapor pressure was defined as the only vapor–liquid equilibrium (VLE) pressure of a pure substance at fixed temperature.

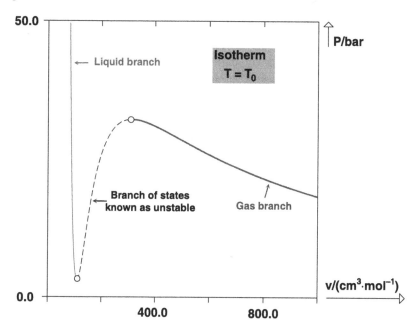

Figure 2.13. *Identification of the liquid, gas and unstable branches on a Van der Waals isotherm*

Since VLE implies simultaneous presence of a gas phase and of a liquid phase in equilibrium, the saturated vapor pressure $P^{sat}(T)$ necessarily belongs to the pressure range in which the equation of state has three possible roots (a liquid root, a gas root and an unstable root), defined in Figure 2.12. The presence of this unique VLE pressure at fixed temperature T_0 is illustrated in Figure 2.14. The associated liquid root $v_L^{sat}(T_0)$ defines the *saturated liquid* state and the associated gas root $v_G^{sat}(T_0)$, the *saturated gas* state (or *saturated vapor*). For all pressures greater than the saturated vapor pressure, the stable state of the fluid is therefore liquid. Similarly, for all pressures lower than the saturated vapor pressure, the stable state of the fluid is gaseous.

Thus, for all pressures between the saturated vapor pressure and the local maximum pressure, the equation of state has three roots: the liquid root is stable, the intermediate root is unstable and the gas root is declared *metastable*: it does not violate the stability criterion but it is not the most stable. Similarly, for all pressures between the saturated vapor pressure and the local minimum, the equation of state has three roots: the gas root is stable, the intermediate root is unstable and the liquid root is *metastable*. The stable and metastable branches of the isotherm are indicated in Figure 2.14.

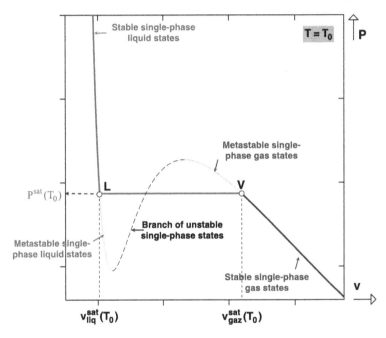

Figure 2.14. *Identification of the liquid, gas and unstable branches on a Van der Waals isotherm. The horizontal line LV defines the diphasic region of the isotherm (often known as "vapor–liquid equilibrium threshold")*

In summary: when the equation of state has a single root v at fixed (T,P), there is no ambiguity: the (T,P,v) state can be declared stable (i.e. this is what would be observed by experimental measurements). It is either a liquid (if $P > P^{sat}(T)$) or a gas (if $P < P^{sat}(T)$). On the other hand, when the equation has three roots at fixed (T,P) (see Figure 2.12), determination of the stable state or states is more complex. In this case, three situations are possible:

– if $P > P^{sat}(T)$, the liquid root is stable and the gas root is metastable;

– if $P < P^{sat}(T)$, the gas root is stable and the liquid root is metastable;

– if $P = P^{sat}(T)$, the liquid and gas roots are stable.

Intermediary roots are always (mechanically) unstable.

2.6.3. Graphical determination of the saturated vapor pressure by the Maxwell equal area rule

The reasoning is shown in Figure 2.15.

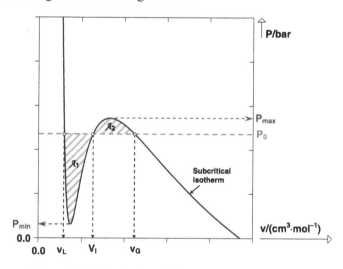

Figure 2.15. *Illustration of the Maxwell equal area rule*

Let P_0 be any pressure in the range $[P_{min}, P_{max}]$, in other words delineating two areas A_1 and A_2 on a subcritical isotherm:

$$A_1 = P_0(v_l - v_L) - \int_{v_L}^{v_l} Pdv \; (>0) \quad and \quad A_2 = \int_{v_l}^{v_G} Pdv - P_0(v_G - v_l) \; (>0)$$

$$[2.48]$$

Thus:

$$
\begin{aligned}
\mathcal{A}_2 - \mathcal{A}_1 &= \int_{v_L}^{v_I} P dv + \int_{v_I}^{v_G} P dv - P_0(v_G - v_I) - P_0(v_I - v_L) \\
&= \underbrace{-\int_{v_G}^{v_L} P dv}_{a(T,v_L)-a(T,v_G)} + P_0(v_L - v_G) = g(T,v_L) - g(T,v_G)
\end{aligned}
\qquad [2.49]
$$

The VLE condition imposing $g(T,v_L) = g(T,v_G)$ (see equation [1.6]) then leads to:

$$
\mathcal{A}_1 = \mathcal{A}_2 \text{ at VLE, i.e. when } \begin{cases} P_0 = P^{sat}(T) \\ v_L = v_L^{sat}(T) \\ v_G = v_G^{sat}(T) \end{cases} \qquad [2.50]
$$

In other words, **the pressure $P^{sat}(T)$ delineates two equal areas on the isotherm**, as illustrated in Figure 2.16.

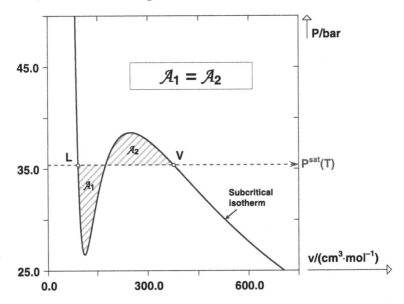

Figure 2.16. *Illustration of the Maxwell equal area rule (continued): when the pressure P_0 is such that the areas $\mathcal{A}_1$ and $\mathcal{A}_2$ are equal, P_0 is equal to the saturated vapor pressure asserted by the cubic equation of state*

2.6.4. *Plot of the stable portions of an isotherm from an equation of state*

We begin by resolving VLE at a fixed temperature (i.e. we calculate $P^{sat}(T)$, $v_L^{sat}(T)$, $v_G^{sat}(T)$).

We calculate the stable single-phase liquid and gas parts of the isotherm by direct application of the equation of state (to do this, we construct a table similar to the one in Figure 2.17).

The graph for the stable portions of the isotherm leads to the graph on the right-hand side of Figure 2.18.

v	$P^*(T,v) = \dfrac{RT}{v-b} - \dfrac{a(T)}{(v-br_1)(v-br_2)}$
b	$+\infty$
$\vdots$	$\vdots$
$v_L^{sat}(T)$	$P^{sat}(T)$
$v_G^{sat}(T)$	$P^{sat}(T)$
$\vdots$	$\vdots$
$+\infty$	0

Figure 2.17. *Table of data to obtain with the objective of constructing an isotherm from an equation of state*

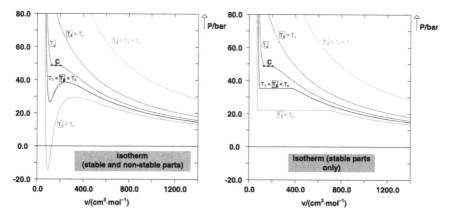

Figure 2.18. *Left: Van der Waals isotherms (including metastable and unstable parts). Right: isotherms only made up of stable states*

2.6.5. *Calculation of the vaporization properties using an equation of state*

Since calculation of the properties $P^{sat}(T)$, $v_L^{sat}(T)$ and $v_G^{sat}(T)$ has been carried out in previous sections, calculation of the vaporization properties is immediate:

$$\Delta_{vap}X(T) = x(T,v_G^{sat}) - x(T,v_L^{sat})$$
$$= \underset{\substack{T \to T \\ v_L^{sat} \to v_G^{sat}}}{\Delta x^{\bullet}} + \underset{\substack{T \to T \\ v_L^{sat} \to v_G^{sat}}}{\Delta x^{res-TV}} \qquad [2.51]$$

The properties u, h, c_P and c_v of a perfect gas only depend on the temperature, and the associated perfect gas terms are equal to zero (see equation [2.46]). We therefore have:

$$\begin{cases}
\Delta_{vap}H(T) = \underset{\substack{T \to T \\ v_L^{sat} \to v_G^{sat}}}{\Delta h^{res-TV}} = h^{res-TV}(T,v_G^{sat}) - h^{res-TV}(T,v_L^{sat}) \\[4mm]
\Delta_{vap}U(T) = \underset{\substack{T \to T \\ v_L^{sat} \to v_G^{sat}}}{\Delta u^{res-TV}} = u^{res-TV}(T,v_G^{sat}) - u^{res-TV}(T,v_L^{sat}) \\[4mm]
\Delta_{vap}C_P(T) = \underset{\substack{T \to T \\ v_L^{sat} \to v_G^{sat}}}{\Delta c_P^{res-TV}} = c_P^{res-TV}(T,v_G^{sat}) - c_P^{res-TV}(T,v_L^{sat}) \\[4mm]
\Delta_{vap}C_v(T) = \underset{\substack{T \to T \\ v_L^{sat} \to v_G^{sat}}}{\Delta c_v^{res-TV}} = c_v^{res-TV}(T,v_G^{sat}) - c_v^{res-TV}(T,v_L^{sat})
\end{cases} \qquad [2.52]$$

This is not the case of the molar entropy (see equation [2.46]):

$$\Delta_{vap}S(T) = \underset{\substack{T \to T \\ v_L^{sat} \to v_G^{sat}}}{\Delta s^{\bullet}} + \underset{\substack{T \to T \\ v_L^{sat} \to v_G^{sat}}}{\Delta s^{res-TV}} = R\ln\left(\frac{v_G^{sat}}{v_L^{sat}}\right) + s^{res-TV}(T,v_G^{sat}) - s^{res-TV}(T,v_L^{sat})$$

$$[2.53]$$

As a reminder, expressions for the residual-TV properties are given by equations [2.41] and [2.42].

2.7. Overall summary: criteria for selecting an equation of state for modeling of the thermodynamic properties of a given pure fluid

– Is the selected pressure low (< 5 bar) and is the pure substance gaseous? If yes, choose a perfect gas model.

– **Is the pressure moderate (< 20 bar) and is the pure substance gaseous?** If yes, the truncated virial equation can be used (on the condition that we have a suitable expression for coefficient $B(T)$).

– Otherwise, is there a specific equation of state for the pure substance in question (Span–Wagner, GERG, BWR, etc.)? If yes, this model will provide the maximum accuracy that can be obtained from an equation of state.

– **Otherwise, is the pure substance slightly to averagely polar and non-associating?** If yes, cubic equations of state are the most appropriate. **Warning:** to put these equations into practice, you need the experimental values for $T_{c,exp}$ and $P_{c,exp}$. The function $\alpha(T)$ plays an essential role (select it carefully). A generalized function (e.g. Soave) can be used but requires knowledge of the acentric factor ω_{exp}. A function involving parameters that are fitted specifically to the pure substance in question can increase the accuracy of the model, but it is advised to ensure in advance that the parameters of the specific function $\alpha(T)$ are readily available. A *volume translation* allows the estimations of the liquid densities to be significantly improved.

– **Otherwise, is the pure substance highly polar and associating?** In this case, envisage a SAFT equation of state (including an association term). Check in advance the availability of the parameters for the SAFT equation for pure substances.

3

Low-Pressure Vapor–Liquid and Liquid–Liquid Equilibria of Binary Systems: Activity-Coefficient Models

---

3.1. Introduction

As a first approach, "low pressure" can be interpreted as a pressure close to atmospheric pressure. In practice, the "*low pressure*" hypothesis is generally applied up to 10 bar (a more precise explanation of this expression is given in the next paragraph).

Consider a closed system containing p components of known global composition, represented by the vector $\mathbf{z} = (z_1, ..., z_p)$ of the molar fractions of the components ($z_i = n_i / \sum_k n_k$). If this system is placed at a temperature T and under a pressure P, it is then likely to contain several fluid phases in equilibrium: mainly vapor–liquid equilibrium (VLE), liquid–liquid equilibrium (LLE), vapor–liquid–liquid equilibrium (VLLE). These equilibria and the notations used to designate molar fractions in the various phases are summarized in Figure 3.1.

In this chapter, we aim to present the "activity coefficient" thermodynamic models that allow this type of equilibrium to be calculated, and to explain how to use them.

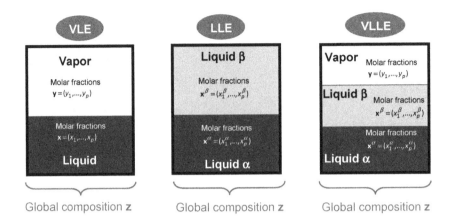

Figure 3.1. *Representation of the main phase equilibria of binary systems at low pressure*

3.2. Classification of fluid-phase behaviors of binary systems at low pressure and low temperature

To precisely explain what we mean by "low pressure" and "low temperature", we use the diagram in Figure 3.2.

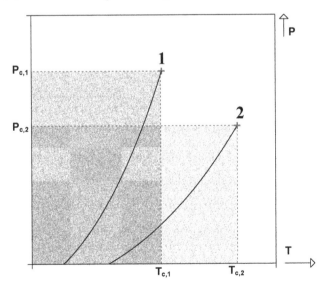

Figure 3.2. *Definition of "low pressure" and "low temperature" regions. For a color version of the figure, see www.iste.co.uk/jaubert/thermodynamic.zip*

All binary systems are formed by two compounds denoted as "1" and "2". The vaporization curves of these compounds are represented in Figure 3.2 (there are several possible configurations of the relative positions of these vaporization curves; the one given in Figure 3.2 is the most common).

The range of *low temperatures* and *low pressures* is defined as the temperature and pressure range in which the binary system does not exhibit vapor–liquid critical points. This range generally corresponds to the range where the two pure substances can exist simultaneously in VLE. Reminding ourselves that a given pure substance can be in VLE if $P < P_c$ and $T < T_c$, the range of low temperatures and low pressures is as a general rule defined by:

$$\begin{cases} P < \min\left\{ P_{c,1} \, ; P_{c,2} \right\} \\ T < \min\left\{ T_{c,1} \, ; T_{c,2} \right\} \end{cases} \qquad [3.1]$$

This range corresponds to the gray-blue part of Figure 3.2. To be rigorous, it is necessary to apply a security margin to equation [3.1] because binary critical points can exist at pressures slightly lower than $\min\{P_{c,1}$; $P_{c,2}\}$ and at temperatures slightly lower than $\min\left\{T_{c,1} \, ; T_{c,2}\right\}$.

3.2.1. Conventions for representation of isobaric and isothermal phase diagrams of binary systems

From here on, we denote the most volatile component as "1" and the heaviest component as "2". In other words, at a fixed temperature: $P_1^{sat}(T) > P_2^{sat}(T)$ and at a fixed pressure: $T_{éb,1}(P) < T_{éb,2}(P)$. In order to represent the VLE of the binary systems graphically, we usually construct an isothermal or isobaric phase diagram.

In an isothermal phase diagram (denoted as "*Pxy*" diagram), the change of the vapor–liquid equilibrium pressure (P_{VLE}) of the binary system is plotted as a function of x_1 (the molar fraction of component 1 in the liquid phase) and as a function of y_1 (the molar fraction of component 1 in the vapor phase). On the x-axis, x_1 and y_1 vary between 0 (0% of component 1,

100% of component 2) and 1 (100% of component 1, 0% of component 2). We thus obtain a diagram containing two curves (called the bubble-point curve and the dew-point curve respectively), which meet at $x_1 = y_1 = 0$ at the vapor pressure of component 2 and at $x_1 = y_1 = 1$ at the vapor pressure of component 1 at the temperature T for the diagram. Effectively, when the binary system only contains the component 1 ($x_1 = y_1 = 1$), it reaches VLE at a pressure equal to $P_1^{sat}(T)$ and when it only contains the component 2 ($x_1 = y_1 = 0$), it reaches VLE at a pressure equal to $P_2^{sat}(T) < P_1^{sat}(T)$. This allows the points that represent the **pure substances** on the *Pxy* diagram to be identified; they are indicated by the small circular symbols in the right-hand side of Figure 3.3.

Similarly, in an isobaric phase diagram (denoted as "*Txy*" diagram), the changes in the vapor–liquid equilibrium temperature (T_{VLE}) of the binary system are plotted as a function of x_1 and y_1, allowing us to define the bubble and dew curves that meet in $x_1 = y_1 = 0$ at the boiling temperature of the component 2 and in $x_1 = y_1 = 1$ at the boiling temperature of the component 1. An example diagram of this type is schematically represented on the right-hand side of Figure 3.4, and we should note in particular the relative position of points that represent pure substances as $T_{b,1}(P) < T_{b,2}(P)$.

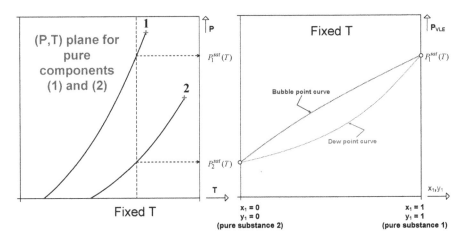

Figure 3.3. *Conventions for representation of a Pxy isothermal phase diagram. For a color version of the figure, see www.iste.co.uk/jaubert/thermodynamic.zip*

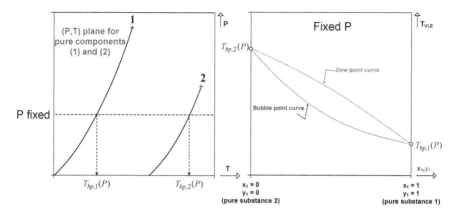

Figure 3.4. *Conventions for representation of a Txy isobaric phase diagram. For a color version of the figure, see www.iste.co.uk/jaubert/thermodynamic.zip*

3.2.2. The four types of vapor–liquid equilibrium (VLE) diagrams for low temperature, low pressure

The following are the four types:

– VLE-1: basic configuration;

– VLE-2: positive homogeneous azeotropy;

– VLE-3: negative homogeneous azeotropy;

– VLE-4: double azeotropy.

We will now give a short description of each of them.

– VLE-1: basic configuration

This case is very common (about 80% of binary systems). *Examples of binary systems in this category:* propane + n-pentane, nitrogen + oxygen, benzene + toluene, CO_2 + H_2S, etc.

Figure 3.5 illustrates this case and demonstrates that in an isothermal diagram, the bubble-point curve is located above the dew-point curve, whereas the opposite is true in an isobaric diagram. This inversion is a simple consequence of the fact that at a high pressure, the binary system is liquid, but at a high temperature, it is gaseous. The domain delineated by the bubble-point and dew-point curves corresponds to the two-phase vapor–liquid domain.

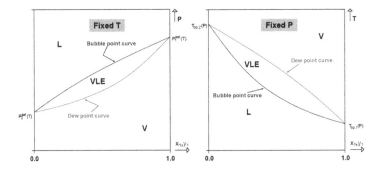

Figure 3.5. *Pxy isothermal diagram and Txy isobaric diagram for the basic type. L: single-phase liquid domain, V: single-phase vapor domain, VLE: two-phase vapor–liquid domain. For a color version of the figure, see www.iste.co.uk/jaubert/ thermodynamic.zip*

NOTE.– The specific case of ideal liquid phases in equilibrium with a perfect gas phase is a sub-category of the "basic configuration" type. The notion of "ideality" is specified in section 3.5.1. It should be remembered that a binary liquid phase is ideal when the molecules A and B that constitute it are of similar sizes and shapes and, in addition, when the energies of the interactions A–A, B–B and A–B are identical (on average). For example, the following liquid systems can be considered ideal: 2-methylpentane + 3-methylpentane, toluene + ethylbenzene, butan-1-ol + butan-2-ol, etc.

For systems of this kind, the bubble-point isotherm is a straight line and the dew-point isotherm is a branch of a hyperbole, as illustrated in Figure 3.6.

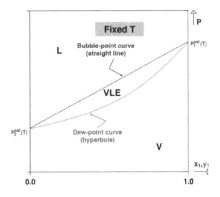

Figure 3.6. *Pxy isothermal diagrams where the liquid phase is ideal and the gas phase is perfect. For a color version of the figure, see www.iste.co.uk/jaubert/thermodynamic.zip*

NOTE.– When the difference in size (i.e. of molecular weight) between the molecules of "basic configuration"-type binary systems increases, the diagrams become deformed, as illustrated in Figure 3.7. Example: propane + n-decane.

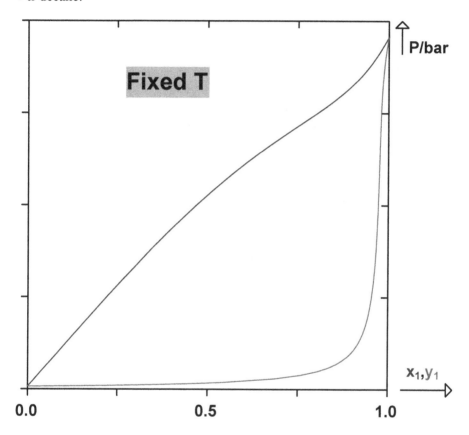

Figure 3.7. *Form of "basic configuration" Pxy diagrams in the case of systems that are asymmetrical in size. For a color version of the figure, see www.iste.co.uk/jaubert/thermodynamic.zip*

– VLE-2: positive homogeneous azeotropy

This phenomenon occurs in approximately 15% of binary systems. *Examples of binary systems in this category:* benzene + cyclohexane, CO_2 + ethane, etc.

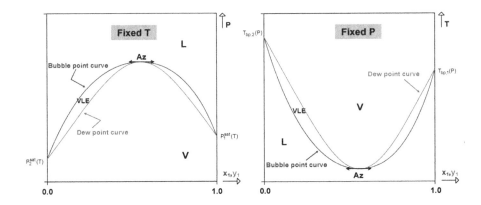

Figure 3.8. *"Positive homogeneous azeotropy" Pxy isothermal and Txy isobaric diagrams. L: single-phase liquid domain, V: single-phase vapor domain, VLE: two-phase vapor–liquid domain. Az: azeotrope. For a color version of the figure, see www.iste.co.uk/jaubert/thermodynamic.zip*

DEFINITION.– An azeotrope can be defined as an extremum (minimum or maximum) that is common to the bubble-point and dew-point curves in a *Pxy* isothermal diagram or in a *Txy* isobaric diagram:

On a *Pxy* diagram:

$$\left(\frac{\partial P_{bubble}}{\partial x_1} \right)_T = \left(\frac{\partial P_{dew}}{\partial y_1} \right)_T = 0$$

On a *Txy* diagram :

$$\left(\frac{\partial T_{bubble}}{\partial x_1} \right)_P = \left(\frac{\partial T_{dew}}{\partial y_1} \right)_P = 0$$

$$\Rightarrow \begin{cases} x_1 = y_1 \\ x_2 = y_2 \end{cases}$$

This condition implies equality of the composition of the liquid and vapor phases in equilibrium.

The azeotrope is said to be **positive** if the bubble-point and dew-point curves have a common maximum in a *Pxy* diagram and correlatively a common minimum in a *Txy* diagram.

– VLE-3: negative homogeneous azeotropy

Much less common, this is the case for about 5% of binary systems. *Example of binary system in this category:* acetone + chloroform.

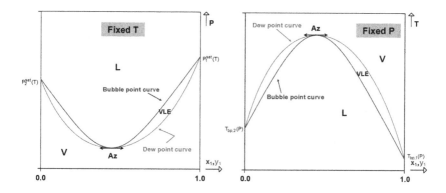

Figure 3.9. *"Negative homogeneous azeotropy" Pxy isothermal and Txy isobaric diagrams. L: single-phase liquid domain, V: single-phase vapor domain, VLE: two-phase vapor–liquid domain. Az: azeotrope. For a color version of the figure, see www.iste.co.uk/jaubert/thermodynamic.zip*

Negative azeotropy behavior is characterized by a common minimum of the bubble-point and dew-point curves in a *Pxy* diagram and correlatively by a common maximum in a *Txy* diagram.

– VLE-4: double azeotropy

This occurs in extremely rare cases. Only about 20 systems presenting this type of behavior have been identified to date. *Example:* benzene + hexafluorobenzene.

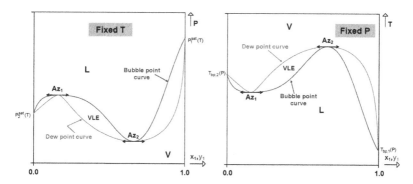

Figure 3.10. *"Double azeotropy" Pxy isothermal and Txy isobaric diagrams. L: single-phase liquid domain, V: single-phase vapor domain, VLE: two-phase vapor–liquid domain. Az$_1$ and Az$_2$: azeotropic points. For a color version of the figure, see www.iste.co.uk/jaubert/thermodynamic.zip*

NOTE.– The positions of the azeotropic points in the *Pxy* diagrams change as the temperature changes (and reciprocally, in *Txy* diagrams, they change as the pressure changes). In the case of double azeotropes, it can sometimes occur that at a certain temperature (and reciprocally at a certain pressure), both azeotropes, one positive and the other negative, merge together to produce a saddle-point azeotrope, as illustrated in Figure 3.11. The azeotropic point is then an inflection point with a horizontal tangent that is common to both the bubble-point and dew-point curves, and the azeotropic pressure is between P_1^{sat} and P_2^{sat}.

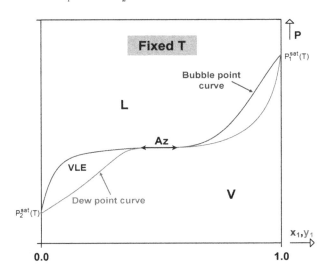

Figure 3.11. *Illustration of a saddle-point azeotrope in a Pxy isothermal diagram. L: single-phase liquid domain, V: single-phase vapor domain, VLE: two-phase vapor–liquid domain. Az: saddle-point azeotrope. For a color version of the figure, see www.iste.co.uk/jaubert/thermodynamic.zip*

3.2.3. *Five types of liquid–liquid equilibrium (LLE) diagrams*

Experimental results show that by mixing two liquid pure substances in appropriate proportions and under certain conditions of temperature and pressure, we do not always obtain a homogeneous liquid solution, but instead two liquid phases of different compositions, denoted as L_α and L_β.

We have then created a **liquid–liquid equilibrium (LLE)**. We also talk about *liquid–liquid demixing*.

Liquid–liquid equilibrium diagrams are generally drawn at a constant pressure and show the variation of the LLE temperature as a function of x_1^α and x_1^β (the molar fraction of component 1 in phases L_α and L_β respectively). We generally denote this as Tx.

– LLE-1:

From a descriptive point of view, the most general case (type I diagram) is the case where the two-phase liquid–liquid domain is closed (see Figure 3.12).

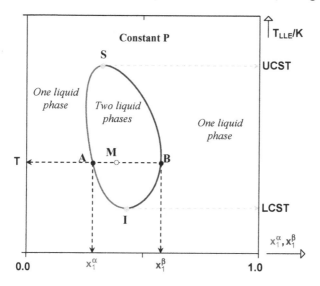

Figure 3.12. *First type of LLE diagram. UCST = upper critical solution temperature, LCST = lower critical solution temperature. For a color version of the figure, see www.iste.co.uk/jaubert/thermodynamic.zip*

Any point M located inside the closed loop, known as the *saturation curve*, is a two-phase point. The composition of the liquid phase rich in component 2 (α phase) can be read from the branch *SAI* and the composition of the liquid phase rich in component 1 (β phase) can be read from the branch *SBI*. The segment *AB* is an equilibrium line. Since the phase diagram is closed, it has **two liquid–liquid critical points**. The highest temperature at which LLE is obtained is known as the *upper critical solution temperature* (UCST). The lowest temperature at which LLE is observed is known as the *lower critical solution temperature* (LCST).

Thus, if the temperature T is between the LCST and the UCST, there is a range of compositions for which an LLE is obtained.

On the other hand, if $T < LCST$ or if $T > UCST$, a homogeneous liquid solution is obtained over the entire composition domain. At the two critical points, the properties of the two liquid phases become identical, in a similar way to what happens at the critical point of a pure substance where the liquid and vapor phases become indistinguishable. Obtaining a type-1 isobaric LLE diagram means that there is no overlap between, on the one hand, the bubble-point curve on the VLE diagram or the liquidus curve on the SLE (solid–liquid equilibrium) diagram, and, on the other hand, the LLE diagram as shown in Figure 3.13.

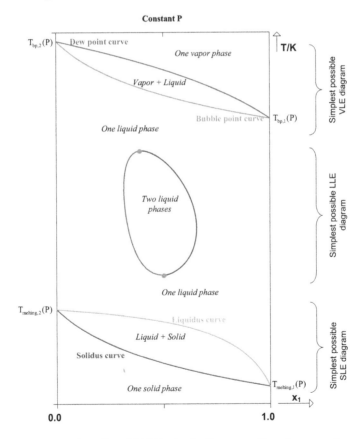

Figure 3.13. *VLE, LLE and SLE isobaric diagrams. The absence of overlap of the various parts produces an LLE diagram of type LLE-1. For a color version of the figure, see www.iste.co.uk/jaubert/thermodynamic.zip*

This behavior is in fact very rare. Only about 4% of systems where liquid–liquid demixing occurs are of type LLE-1. As an example, we can cite the binary systems THF + water and nicotine + water.

When the LLE diagram overlaps with the SLE diagram (the liquidus curve), the lower critical point disappears and we obtain a type 2 LLE diagram, which only has a UCST. At low temperatures, the LLE diagram ends with an SLL three-phase line (the result of the overlap between an LL domain and an LS domain).

If the LLE diagram intercepts the bubble-point curve on a VLE diagram, the upper critical point disappears and we obtain a type 3 LLE diagram, which has only an LCST. At high temperatures, the LLE diagram ends with a VLL three-phase line (consequence of overlap of an LL domain and an LV domain).

If the LLE diagram simultaneously intercepts the VLE diagram and the SLE diagram, no liquid–liquid critical point exists and we obtain a type 4 LLE diagram. This type of diagram is limited at high temperatures by a VLL three-phase line and at low temperatures by an SLL three-phase line. Figures 3.14–3.16 illustrate LLE-2, LLE-3 and LLE-4 diagrams.

– LLE-2:

LLE-2 type diagrams have a frequency of about 40%. Among the systems showing this behavior, we can cite, for example: methyl acetate + water, butan-1-al + water, aniline + n-hexane.

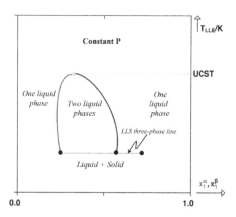

Figure 3.14. *Second type of LLE diagram. For a color version of the figure, see www.iste.co.uk/jaubert/thermodynamic.zip*

– LLE-3:

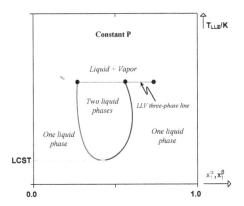

Figure 3.15. *Third type of LLE diagram. For a color version of the figure, see www.iste.co.uk/jaubert/thermodynamic.zip*

The frequency of LLE-3 diagrams is approximately 2%. Among the systems showing this behavior, we can cite the examples: dipropylamine + water, 2,4-dimethylpyridine + water.

– LLE-4:

The frequency of LLE-4-type diagrams is approximately 50%. Among the systems showing this behavior, we can mention, for example: benzene + water, tetrachloromethane + water.

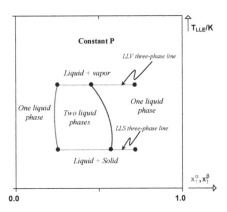

Figure 3.16. *Fourth type of LLE diagram. For a color version of the figure, see www.iste.co.uk/jaubert/thermodynamic.zip*

– LLE-5:

There is also a fifth type of diagram which, in the same way as the LLE-1 diagram, has two liquid–liquid critical points but for which the two-phase domain is not closed. The upper critical solution temperature (UCST) is then below the lower critical solution temperature (LCST).

The occurrence of type 5 LLE diagrams is approximately 4%. Among the rare systems with this behavior, we can cite, for example: sulfur + aromatic compound (benzene, toluene, p-xylene, triphenylmethane), polystyrene + acetone.

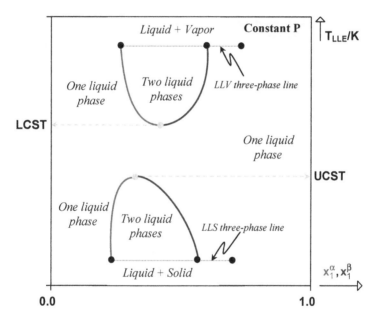

Figure 3.17. *Fifth type of LLE diagram. For a color version of the figure, see www.iste.co.uk/jaubert/thermodynamic.zip*

To conclude this section on LLE diagrams, we will mention that at low pressures and low temperatures, we can reasonably apply the assumption of **incompressible liquid**. This hypothesis means that the LLE isobaric diagram remains identical, regardless of the pressure at which the diagram is drawn. In other words, an isothermal LLE diagram in the plane pressure–molar fraction of component 1 is made of two straight vertical lines, as illustrated in Figure 3.18.

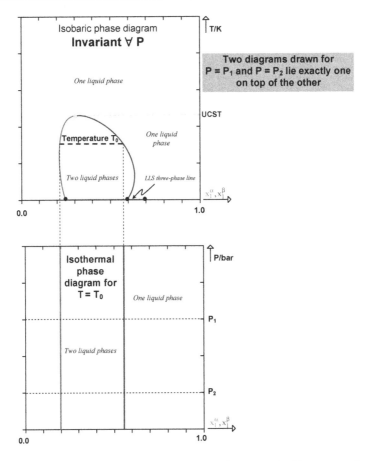

Figure 3.18. *LLE diagrams in the Tx and Px planes, under the assumption that the liquid phases are incompressible. For a color version of the figure, see www.iste.co.uk/jaubert/thermodynamic.zip*

3.2.4. *Fluid-phase diagrams resulting from the overlap of LLE and VLE domains. Liquid–liquid–vapor equilibria (VLLE)*

We have mentioned previously that the LLE diagram can interfere with the bubble-point curve on the VLE diagram. When this occurs, it leads to a **three-phase** (liquid–liquid–vapor) **equilibrium.**

Four different situations can occur depending on whether the VLE diagram exhibits a basic configuration form, a positive azeotrope or a negative azeotrope. In this paragraph, we will give an overview of the following four cases:

– **VLLE-1:** overlap of a basic-configuration VLE diagram and the LLE diagram;

– **VLLE-2:** overlap of a VLE diagram with a positive azeotrope, and the LLE diagram:

  - **VLLE-2a:** case where the overlap does not include the azeotropic composition;

  - **VLLE-2b:** case where the overlap includes the azeotropic composition;

– **VLLE-3:** overlap of a VLE diagram with a negative azeotrope, and the LLE diagram.

VLLE-1: overlap of the LLE diagram and a basic-configuration VLE diagram

We are interested first and foremost in a *Txy* isobaric projection. If the pressure is high enough, the VLE and LLE domains will not overlap.

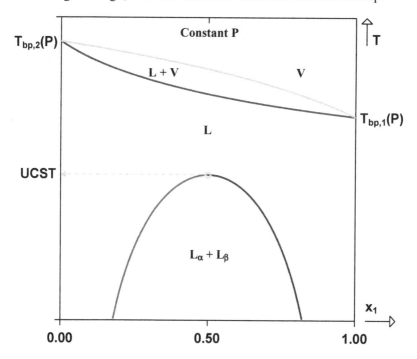

Figure 3.19. *Example of LLE and VLE diagrams that co-exist without overlapping. For a color version of the figure, see www.iste.co.uk/jaubert/thermodynamic.zip*

On the other hand, when the pressure is reduced, the LLE diagram (very low sensitivity to pressure) hardly moves, whereas the VLE diagram will move downwards ($T_{bp,1}(P)$ and $T_{bp,2}(P)$ reducing as pressure reduces) and will interfere with the LLE diagram to give a vapor–liquid–liquid equilibrium of the type represented on the right-hand side of Figure 3.20.

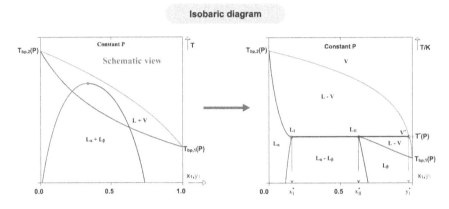

Figure 3.20. *Isobaric projection of VLLE-1 behavior. Left: simplified diagram representing the overlap of the VLE and LLE domains. Right: actual diagram obtained (result of the overlap). For a color version of the figure, see www.iste.co. uk/jaubert/thermodynamic.zip*

T^* is the three-phase temperature at the constant pressure P to which the diagram corresponds (T^* depends on P). At T^*, we therefore have three phases in equilibrium: two liquids (points L_I and L_{II} on the right-hand side of Figure 3.20) and one vapor (point V^* on the same diagram). The three phases are at the same pressure and at the same temperature (they are therefore on the same horizontal line in a *Txy* diagram). The segment $L_I L_{II} V^*$ is a three-phase segment. All points located on this segment represent a fluid in the form of three phases in equilibrium. The composition of the three phases can be read directly from the x-axis of the *Txy* isobaric diagram. On the other hand, this diagram does not allow us to determine the proportion of the three phases in equilibrium.

We will now focus on the isothermal representation of this same phenomenon. As previously explained, assuming the liquid phases are incompressible, an LLE isothermal diagram is represented by two straight vertical lines. The overlap of an LLE isothermal diagram with a VLE isothermal diagram is represented on the left-hand side of Figure 3.21. The result of this intersection is displayed on the right-hand side of this same figure.

Isothermal diagram

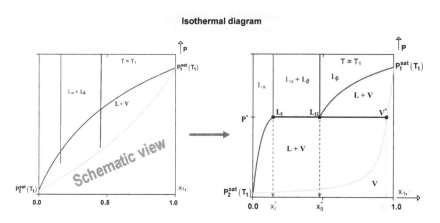

Figure 3.21. *Isothermal projection of VLLE-1 behavior. Left: simplified diagram representing the overlap of the VLE and LLE domains. Right: actual diagram obtained (result of the overlap). For a color version of the figure, see www.iste.co.uk/ jaubert/thermodynamic.zip*

VLLE-2: overlap of a VLE diagram with a positive azeotrope, and the LLE diagram

The same process as above can be applied, but two cases must be distinguished depending on whether the overlap between the LLE diagram and the VLE diagram includes the azeotropic composition or not.

VLLE-2a: overlap of the LLE diagram with a VLE diagram that has a positive azeotrope (the overlap of the two diagrams does not encompass the azeotropic composition)

The corresponding isobaric and isothermal projections are indicated in Figures 3.22 and 3.23.

VLLE-2b: overlap of the LLE diagram with a VLE diagram that has a positive azeotrope (the azeotropic composition is included in the overlap)

The corresponding isobaric and isothermal projections are given in Figures 3.24 and 3.25.

In this case, at the three-phase temperature T^*, we have: $x_I^* \leq y_1^* \leq x_{II}^*$. The bubble-point and dew-point curves touch each other at the point V^*, which is on the three-phase line. This point of contact between the bubble-point and dew-point curves (V^*) defines a **heterogeneous azeotrope** (or **heteroazeotrope**).

Isobaric diagram

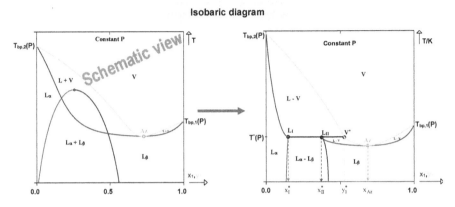

Figure 3.22. *Isobaric projection of VLLE-2a behavior. Left: simplified diagram representing the overlap of the VLE and LLE domains. Right: actual diagram obtained (result of the overlap). For a color version of the figure, see www.iste.co.uk/ jaubert/thermodynamic.zip*

Isothermal diagram

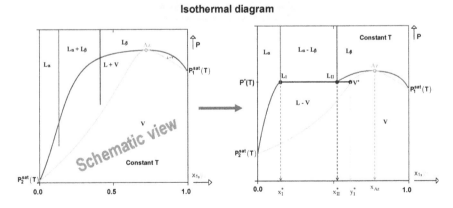

Figure 3.23. *Isothermal projection of VLLE-2a behavior. Left: simplified diagram representing the overlap of VLE and LLE domains. Right: actual diagram obtained (result of the overlap). For a color version of the figure, see www.iste.co.uk/jaubert/ thermodynamic.zip*

Isobaric diagram

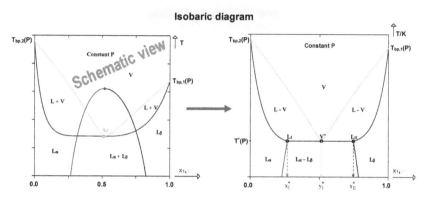

Figure 3.24. *Isobaric projection of VLLE-2b behavior. Left: simplified diagram representing the overlap of the VLE and LLE domains. Right: actual diagram obtained (result of the overlap). For a color version of the figure, see www.iste.co.uk/ jaubert/thermodynamic.zip*

Isothermal diagram

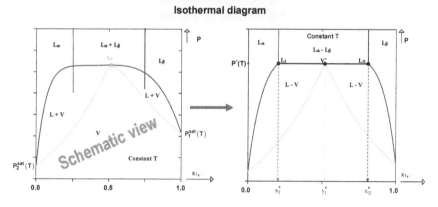

Figure 3.25. *Isothermal projection of VLLE-2b behavior. Left: simplified diagram representing the overlap of the VLE and LLE domains. Right: actual diagram obtained (result of the overlap). For a color version of the figure, see www.iste. co.uk/jaubert/thermodynamic.zip*

NOTE.– Figure 3.26 proposes a comparison between the cases of homogeneous and heterogeneous azeotropic phenomena. For the first case, the liquid phase and the gas phase have the same composition: for any component i, we have $x_i = y_i$. For the second case, for a mixture with global composition that is equal to the heteroazeotropic composition ($z_i = y_i^*$), by mixing the two liquid phases in equilibrium with vapor of composition y_i^*, we would obtain a liquid phase with the same global composition as the gas phase at the heteroazeotropic point.

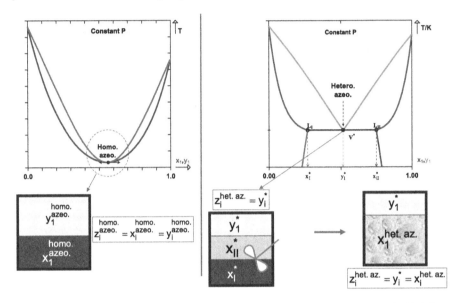

Figure 3.26. *Comparison of homoazeotropic (left) and heteroazeotropic (right) phenomena. For a color version of the figure, see www.iste.co.uk/jaubert/thermodynamic.zip*

– VLLE-3: overlap of an LLE diagram with a VLE diagram that has a negative azeotrope.

This situation is extremely rare. Without dwelling too much on this case, we will all mention the fact that a negative azeotrope is generally associated with a negative molar excess Gibbs energy $g^E < 0$ (g^E denotes the difference between the molar Gibbs energy of the solution and that of an ideal liquid solution; we will return later on to the meaning of this quantity). Yet, when $g^E < 0$, the liquid solution is particularly stable (more stable than the corresponding ideal solution) and therefore does not tend to give rise to liquid–liquid demixing. The binary system triethylamine(1) + ethanoic acid(2) is however an example of this (see Figures 3.27 and 3.28).

In this case, the overlap of the two diagrams cannot include the azeotropic composition, and the only possible case is presented below.

Isobaric diagram

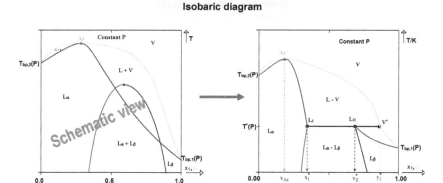

Figure 3.27. *Isobaric projection of VLLE-3 behavior. Left: simplified diagram representing the overlap of the VLE and LLE domains. Right: actual diagram obtained (result of the overlap). For a color version of the figure, see www.iste.co.uk/ jaubert/thermodynamic.zip*

Isothermal diagram

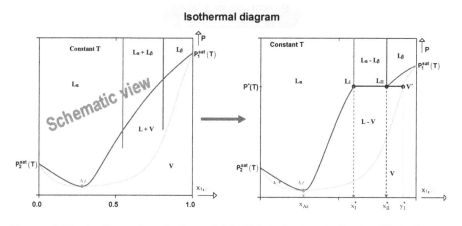

Figure 3.28. *Isothermal projection of VLLE-3 behavior. Left: simplified diagram representing the overlap of the VLE and LLE domains. Right: actual diagram obtained (result of the overlap). For a color version of the figure, see www.iste.co.uk/ jaubert/thermodynamic.zip*

Now that we have summarized the various types of fluid phase behaviors of binary systems at low temperature and low pressure, we will examine how they can be modeled (and to a lesser extent, calculated).

3.3. Condition of equilibrium between fluid phases of binary systems

Based on the same principle as in the cases of pure substances, a multi-phase closed system with uniform phases is said to be in internal equilibrium if and only if:

The system is in **thermal equilibrium**, in other words, there is no exchange of heat between the phases in the system. This condition will be fulfilled **if all the phases have the same temperature**.

The system is in **mechanical equilibrium**, in other words, there is no exchange of pressure–volume work between the phases in the system. This condition will be fulfilled **if all the phases have the same pressure**.

The system is in **chemical equilibrium**, in other words, there is no exchange of matter between the phases in the system. This condition will be fulfilled **if the chemical potentials of each of the components are the same in all the phases of the system**.

Let us apply the condition to a two-phase system containing p components, as shown in the simplified diagram in Figure 3.29.

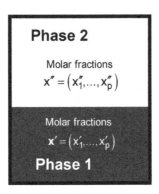

Figure 3.29. *Simplified diagram of a multicomponent two-phase system*

$$\begin{cases} T_{phase\,1} = T_{phase\,2} & (denoted\ T) \\ P_{phase\,1} = P_{phase\,2} & (denoted\ P) \\ \mu_{i,\,phase\,1} = \mu_{i,\,phase\,2} & \forall\ constituent\ i \end{cases} \qquad [3.2]$$

Now let us apply the condition of equilibrium between phases to a binary system in VLE (see Figure 3.30).

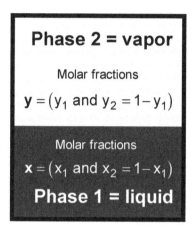

Figure 3.30. *Simplified diagram of a binary system in VLE*

$$\begin{cases} T_{liquid} = T_{gas} = T \\ P_{liquid} = P_{gas} = P \\ \mu_{1,liquid}(T,P,\mathbf{x}) = \mu_{1,gas}(T,P,\mathbf{y}) \\ \mu_{2,liquid}(T,P,\mathbf{x}) = \mu_{2,gas}(T,P,\mathbf{y}) \end{cases} \qquad [3.3]$$

3.4. Vapor–liquid equilibrium relationship at low pressure

Starting with the equality $\mu_{i,liquid} = \mu_{i,gas}$ (valid for the components $i=1$ and $i=2$ of the binary system), we can demonstrate the **VLE relationship at low pressure** that must be viewed as a practical form of the condition of equilibrium between fluid phases. Here, *practical* means that involved quantities are available from experimental results. For a system with p components, the low-pressure VLE relationship is:

$$P \cdot y_i \cdot C_i(T,p,y) = P_i^{sat}(T) \cdot x_i \cdot \gamma_i(T,P,x) \quad \text{for} \quad i = \{1;...;p\} \qquad [3.4]$$

with:

– T and P, the temperature and pressure of the system in VLE respectively;

– x_i and y_i, the molar fraction of component i in the liquid and vapor phases respectively;

– $P_i^{sat}(T)$, the saturated vapor pressure of pure component i at temperature T;

– $\gamma_i(T,P,x)$, the **activity coefficient** of component i in the **liquid phase**. It is essential to estimate this coefficient in order to correctly represent the behavior of the fluid phases. We will detail the models used to estimate these coefficients in the next parts of this chapter.

The function $C_i(T,P,y)$ is given by:

$$C_i\left(T,P,y\right)=\frac{\varphi_{i,vap}\left(T,P,y\right)}{\varphi_i^*\left(T,P_i^{sat}\left(T\right)\right)}\cdot\exp\left(\underbrace{\frac{1}{R\cdot T}\int_P^{P_i^{sat}(T)}v_{i,liq}^*\left(T,P\right)\cdot dP}_{\text{Poynting factor}}\right) \qquad [3.5]$$

with:

– $\varphi_{i,vap}$ and φ_i^*, the fugacity coefficients of component i in the gas phase of the binary system at (T,P,y) on the one hand, and in a pure gas phase at $\left(T,P_i^{sat}\left(T\right)\right)$; they are estimated using an equation of state for gases;

– $v_{i,liq}^*$, the molar volume of the pure liquid component i, which can be estimated from a correlation.

This low-pressure approach, which consists of representing the gas phase by an equation of state and the liquid phase by an activity-coefficient model, is known as the **gamma-phi method** ("$\gamma-\varphi$"). The reason for this notation is that the letter gamma is generally used to designate the activity coefficients (which are produced by activity-coefficient models), whereas the letter phi is used for fugacity coefficients (which result from equations of state).

Applied to a binary system, equation [3.4] is written as:

$$\begin{cases} \mu_{1,liquid}(T,P,x) = \mu_{1,gas}(T,P,y) \\ \mu_{2,liquid}(T,P,x) = \mu_{2,gas}(T,P,y) \end{cases}$$

$$\Updownarrow \qquad\qquad\qquad\qquad [3.6]$$

$$\begin{cases} P \cdot y_1 \cdot C_1(T,P,y) = P_1^{sat}(T) \cdot x_1 \cdot \gamma_1(T,P,x) \\ P \cdot \underbrace{y_2}_{1-y_1} \cdot C_2(T,P,y) = P_2^{sat}(T) \cdot \underbrace{x_2}_{1-x_1} \cdot \gamma_2(T,P,x) \end{cases}$$

We will now look at estimating the various quantities that are involved in equation [3.6] and that we have not described so far in this book.

We give a reminder that the models for the saturated vapor pressure of pure substances (Clapeyron, Antoine, Wagner, etc.) are discussed in sections 1.3.1 and 1.3.2.

3.4.1. *Estimation of the fugacity coefficient of a pure substance, present in the expression of function C_i*

By definition:

$$R \cdot T \cdot \ln \varphi^* \underset{def}{=} g^{res-TP}(T,P) = g^*(T,P) - g^\bullet(T,P) \qquad [3.7]$$

This definition leads to the following for a perfect gas: $\begin{cases} \varphi^\bullet = 1 \\ \ln \varphi^\bullet = 0 \end{cases}$.

This coefficient is calculated using an equation of state for pure substances. We can show that in the case of an **equation of state that is explicit in volume** (the virial equation, for example), the fugacity coefficient is given by:

$$\ln \varphi^*(T,P) = \frac{1}{R \cdot T} \cdot \int_0^P \left[\underbrace{v^*(T,P)}_{\text{Equation of state}} - \frac{R \cdot T}{P} \right] \cdot dP \qquad [3.8]$$

On the other hand, in the case of an **equation of state explicit in pressure** (e.g. the Van der Waals, SRK, PR, SAFT equations), the fugacity coefficient is given by:

$$\ln \varphi^*\left(T,v\right) = \frac{\overbrace{P^*\left(T,v\right)}^{\text{Equation of state}} \cdot v}{R\cdot T} - 1 - \ln\left[\frac{\overbrace{P^*\left(T,v\right)}^{\text{Equation of state}} \cdot v}{R\cdot T}\right]$$

$$-\int_{+\infty}^{v}\left[\frac{\overbrace{P^*\left(T,v\right)}^{\text{Equation of state}}}{R\cdot T} - \frac{1}{v}\right]\cdot dv$$ [3.9]

3.4.2. Estimation of the fugacity coefficient of a component in a mixture, present in the expression of function C_i

By definition:

$$R\cdot T\cdot \ln \varphi_{i,vap} \underset{def}{=} \overline{g}_i^{res-T,P}\left(T,P,y\right) = \overline{g}_i\left(T,P,y\right) - \overline{g}_i^*\left(T,P,y\right)$$ [3.10]

Using this definition, we can deduce that for a mixture of perfect gases $\varphi_i^* = 1$ (and $\ln \varphi_i^* = 0$). This coefficient is calculated using an *equation of state for mixtures*. We can show that in the case of an **equation of state explicit in volume**, the fugacity coefficient is given by:

$$\ln \varphi_i^{vap}\left(T,P,y\right) = \frac{1}{R\cdot T}\int_0^P\left[\left(\frac{\partial \overbrace{V(T,P,n)}^{\text{Equation of state}}}{\partial n_i}\right)_{T,P,n_{j\neq i}} - \frac{R\cdot T}{P}\right]\cdot dP$$ [3.11]

On the other hand, in the case of an **equation of state explicit in pressure**, the fugacity coefficient is given by:

$$
\ln \varphi_i (T, v, y) = -\ln \left[\frac{\overbrace{P(T, v, y) \cdot v}^{\text{Equation of state}}}{R \cdot T} \right]
$$

$$
- \int_{+\infty}^{V} \left[\frac{1}{RT} \left[\frac{\overbrace{\partial P(T, V, n)}^{\text{Equation of state}}}{\partial n_i} \right]_{T, V, n_{j \neq i}} - \frac{1}{V} \right] \cdot dV
$$

[3.12]

3.4.3. Estimation of the Poynting factor, present in the expression of the function C_i

This factor (denoted as $F_{P,i}$) is introduced by equation [3.5]. It includes $v^*_{i,liq}(T, P)$, the molar volume of the pure substance i at (T, P). Applying the hypothesis of an **incompressible liquid** (generally valid at a low to moderate pressure, in other words up to about 20 bar), we then suppose that the molar volume of a liquid does not depend on the pressure. Thus:

$$
F_{P,i} = \exp \left(\frac{1}{R \cdot T} \int_{P}^{P_i^{sat}(T)} v^*_{i,liq}(T, P) \cdot dP \right) \approx \exp \left[\frac{v^*_{i,liq}(T)}{R \cdot T} \left[P_i^{sat}(T) - P \right] \right]
$$

[3.13]

In section 1.4, we have seen some models that can be used to describe the molar volumes of pure substances.

3.4.4. Estimation of the C_i term

In most cases, calculation of the C_i terms in equation [3.6] can be greatly simplified. Effectively, at low pressure, the perfect-gas assumption is realistic and we can assert that:

$$
\varphi_i^{vap}(T, P, y) = \varphi_i^*(T, P_i^{sat}(T)) = 1,0 \quad \Rightarrow \quad \frac{\varphi_{i,vap}(T, P, y)}{\varphi_i^*(T, P_i^{sat}(T))} = 1 \quad [3.14]
$$

At moderate pressure, the fugacity coefficients diverge from the value of 1:

$$\begin{cases} \varphi_i^{vap}\left(T,P,y\right) \neq 1,0 \\ and \\ \varphi_i^*\left(T,P_i^{sat}\left(T\right)\right) \neq 1,0 \end{cases} \quad [3.15]$$

because the perfect-gas assumption is no longer applicable. However, their ratio remains approximately 1:

$$\frac{\varphi_{i,vap}\left(T,P,\mathbf{y}\right)}{\varphi_i^*\left(T,P_i^{sat}\left(T\right)\right)} \approx 1 \quad [3.16]$$

Concerning the Poynting factor, we can generally consider it equal to 1 at a moderate pressure. Common values for the various terms in equation [3.13] are:

$$\begin{cases} T \approx 300 \text{ K} \\ v_{liq}^*\left(T\right) \approx 50 \text{ cm}^3 \cdot \text{mol}^{-1} \\ P_i^{sat}\left(T\right) - P \approx 1 \text{ bar} \end{cases}$$

Applying [3.13], we would then have:

$$F_{P,i} = \exp\left[\frac{\left(50 \cdot 10^{-6} \text{ m}^3 \cdot \text{mol}^{-1}\right) \times \left(10^5 \text{ Pa}\right)}{\left(8.314 \text{ J} \cdot \text{mol}^{-1} \cdot \text{K}^{-1}\right) \times \left(300 \text{ K}\right)}\right] = 1.002 \,.$$

In summary, the factor C_i can generally be considered equal to 1 as long as the pressure remains moderate (< 20 bar).

3.5. Activity coefficients: definition and models

3.5.1. *Ideal solution*

An **ideal solution** is a liquid solution whose energy is independent of the configuration, i.e. of the distribution of the molecules with respect to each other. Strictly speaking, an ideal solution could only be observed if the molecules of the various components were identical and nevertheless distinguishable.

In practice, we could consider that a liquid solution is ideal if the various molecules present in the solution have only small differences in terms of:

$$\begin{cases} \text{size,} \\ \text{shape,} \\ \text{and molecular interactions.} \end{cases}$$

Examples of real solutions that are as close as possible to ideal solutions include a mixture of water (H_2O) and heavy water (D_2O). Two neighboring terms in a homologous series of compounds also give nearly ideal solutions. As an example, we can cite the mixtures:

– n-heptane + n-octane:

– Cyclohexane + cycloheptane:

– Toluene + ethylbenzene:

– Ethanol + propan-1-ol:

3.5.2. *Excess Gibbs energy and activity coefficients*

Excess Gibbs energy, denoted as G^E, is the difference between the Gibbs energy of a real liquid solution and that of an ideal solution at the same temperature, pressure and composition as the real solution:

$$G^E(T,P,\boldsymbol{n}) = G_{\substack{real\ liquid \\ solution}}(T,P,\boldsymbol{n}) - G_{\substack{ideal \\ solution}}(T,P,\boldsymbol{n}) \qquad [3.17]$$

G^E has the dimension of an energy (can be expressed in joule). It is a **measure of the non-ideal nature of a solution**. In other words, the excess Gibbs energy of an ideal solution is zero. The molar excess Gibbs energy (in $J \cdot mol^{-1}$) is denoted as $g^E(T,P,\boldsymbol{x})$ in the following. From the expression for $g_{\substack{ideal \\ solution}}(T,P,\boldsymbol{x})$ provided by statistical thermodynamics, we obtain:

$$g^E(T,P,\boldsymbol{x}) = g_{\substack{real\ liquid \\ solution}}(T,P,\boldsymbol{x}) - g_{\substack{ideal \\ solution}}(T,P,\boldsymbol{x})$$

$$= g_{\substack{real\ liquid \\ solution}}(T,P,\boldsymbol{x}) - \left[\sum_{i=1}^{P} x_i g_{liquid,i}^{*}(T,P) + RT \sum_{i=1}^{P} x_i \ln x_i \right] \qquad [3.18]$$

The activity coefficient γ_i of a component i in solution is deduced from the expression for the excess Gibbs energy according to:

$$R \cdot T \cdot \ln \gamma_i (T,P,\boldsymbol{x}) \underset{déf}{=} \left(\frac{\partial G^E}{\partial n_i} \right)_{T,P,n_{j \neq i}} \qquad [3.19]$$

In the case of an ideal solution, we have: $\gamma_{\substack{i,ideal \\ solution}} = 1$.

The expression [3.19] is not very practical because activity-coefficient models are presented in the form $g^E(T,P,\boldsymbol{x})$ and not $G^E(T,P,\boldsymbol{n})$. For the case of binary systems, we can give the following expressions, which are much simpler to use:

$$
\begin{cases}
\textbf{For a binary system :} \\[4pt]
\ln \gamma_1 = g^E /(RT) + x_2(\delta_1 - \delta_2) \\[4pt]
\ln \gamma_2 = g^E /(RT) + x_1(\delta_2 - \delta_1) \\[4pt]
\text{with: } \delta_i = \left(\dfrac{\partial g^E /(RT)}{\partial x_i} \right)_{T,x_{j\neq i}}
\end{cases}
\qquad [3.20]
$$

3.5.3. Classification of activity-coefficient models

All models for g^E in the scientific literature presume that the liquid phases are incompressible; in other words, they presume that the pressure has no influence on the value of g^E.

There are essentially two types of activity-coefficient models (also known as "g^E models"):

Purely correlative models: these models do not have a theoretical physical (or chemical) basis; they are mathematical expressions for g^E that are flexible (i.e. they have the ability to fit to many experimental behaviors) and involve adjustable parameters.

Models with a theoretical physical (or chemical) basis: starting by decomposing g^E into an enthalpic (h^E) and an entropic ($-T \cdot s^E$) contribution thanks to the equality:

$$
g^E = h^E - T \cdot s^E , \qquad\qquad [3.21]
$$

many authors have put forward g^E expressions deduced from various theories (Van der Waals theory, Guggenheim quasi-reticular theory, etc.). We will note that the **enthalpic effects** are also known as *attractive effects* or **residual**[1] **contribution** (expressing mainly the effect of the dispersion forces, polarity, hydrogen bonds, etc.), whereas the **entropic effects** are also known as *repulsive effects* or **combinatorial contribution** (expressing mainly the effects of the differences in size and shape between the molecules in the mixture).

---

1 This term "residual" has no relation to the term used to describe the values measuring the difference between the property of a real mixture and the property of a perfect gas.

Figure 3.31 presents a summary of the various g^E models that we will introduce in the following sections.

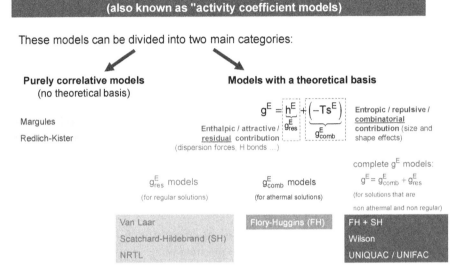

Figure 3.31. *Overview of the two main types of activity-coefficient models. For a color version of the figure, see www.iste.co.uk/jaubert/thermodynamic.zip*

3.5.4. *Purely correlative Margules models*

– Margules activity-coefficient model[2] **with one adjustable parameter (*A*) per binary system:** this is the simplest possible expression.

$$\frac{g^E}{R \cdot T} = A(T) \cdot x_1 \cdot x_2 \qquad [3.22]$$

Expression for the activity coefficients deduced from equations [3.20] and [3.22]:

$$\begin{cases} \ln \gamma_1 = A(T) \cdot x_2^2 \\ \ln \gamma_2 = A(T) \cdot x_1^2 \end{cases} \qquad [3.23]$$

---

2 Max Margules: 1856–1920.

NOTE.– Considering the model $g^E = A \cdot x_1 \cdot x_2$, we obtain an alternative to the Margules model:
$$\begin{cases} \ln \gamma_1 = A \cdot x_2^2 / RT \\ \ln \gamma_2 = A \cdot x_1^2 / RT \end{cases}.$$ This remark remains true for all the models described below.

– Model with two adjustable parameters (A_{12} and A_{21}) per binary system:

$$\begin{cases} \dfrac{g^E}{R \cdot T} = x_1 \cdot x_2 \cdot \left(A_{12} \cdot x_1 + A_{21} \cdot x_2 \right) \\ \text{or} \\ g^E = x_1 \cdot x_2 \cdot \left(A_{12} \cdot x_1 + A_{21} \cdot x_2 \right) \end{cases} \qquad [3.24]$$

Expression for the activity coefficients, deduced from equations [3.20] and [3.24]:

$$\begin{cases} \begin{cases} \ln \gamma_1 = x_2^2 \cdot \left[A_{21} + 2x_1 \cdot \left(A_{12} - A_{21} \right) \right] \\ \ln \gamma_2 = x_1^2 \cdot \left[A_{12} + 2x_2 \cdot \left(A_{21} - A_{12} \right) \right] \end{cases} \\ \text{or} \\ \begin{cases} \ln \gamma_1 = x_2^2 \cdot \left[A_{21} + 2x_1 \cdot \left(A_{12} - A_{21} \right) \right] / (RT) \\ \ln \gamma_2 = x_1^2 \cdot \left[A_{12} + 2x_2 \cdot \left(A_{21} - A_{12} \right) \right] / (RT) \end{cases} \end{cases} \qquad [3.25]$$

– Extension of the Margules models to multicomponent systems that contain more than two compounds:

We then define the molar excess Gibbs energy by:

$$g^E = \sum_{\substack{sum \text{ over all} \\ binary \text{ systems}}} g_{ij}^E \quad \text{with} \quad \begin{cases} g_{ij}^E = A_{ij} x_i x_j & \text{[Margules 1]} \\ \text{or} \\ g_{ij}^E = x_i x_j \left(A_{ij} x_i + A_{ji} x_j \right) & \text{[Margules 2]} \end{cases} \qquad [3.26]$$

We recall that in a system of p components, we can define $p(p-1)/2$ binary systems. We define the total quantity of matter by the letter n. The corresponding excess Gibbs energy is:

$$G^E = n \cdot g^E = \sum_{i=1}^{p} \sum_{j>i}^{p} n \cdot g_{ij}^E$$

$$\text{with} \begin{cases} n \cdot g_{ij}^E = A_{ij} n_i n_j / n & \text{[Margules 1]} \\ \text{or} \\ n \cdot g_{ij}^E = n_i n_j \left(A_{ij} n_i + A_{ji} n_j \right) / n^2 \\ \qquad\qquad\qquad\qquad \text{[Margules 2]} \end{cases} \qquad [3.27]$$

From equation [3.19], we deduce the expression for the activity coefficients:

$$RT \ln \gamma_k = \sum_{i=1}^{p} \sum_{j>i}^{p} \left[\frac{\partial \left(n \cdot g_{ij}^E \right)}{\partial n_k} \right]_{T, n_{\ell \neq k}} \qquad [3.28]$$

3.5.5. Redlich–Kister purely correlative model

– **Model with** $N+1$ **adjustable parameters per binary system** (N **designates the selected degree of the Redlich–Kister polynomial):**

$$\begin{cases} \dfrac{g^E}{RT} = x_1 x_2 \sum_{k=0}^{N} A_k \left(x_1 - x_2 \right)^k \\ \text{or} \\ g^E = x_1 x_2 \sum_{k=0}^{N} A_k \left(x_1 - x_2 \right)^k \end{cases} \qquad [3.29]$$

Expression of the activity coefficients deduced from equations [3.20] and [3.29]:

$$\begin{cases} \begin{cases} \ln \gamma_1 = x_2^2 \sum_{k=0}^{N} \left[2x_1 \left(1+k \right) - 1 \right] A_k \left(x_1 - x_2 \right)^{k-1} \\ \ln \gamma_2 = x_1^2 \sum_{k=0}^{N} \left[1 - 2x_2 \left(1+k \right) \right] A_k \left(x_1 - x_2 \right)^{k-1} \end{cases} \\ \text{or} \\ \begin{cases} \ln \gamma_1 = \dfrac{x_2^2}{RT} \sum_{k=0}^{N} \left[2x_1 \left(1+k \right) - 1 \right] A_k \left(x_1 - x_2 \right)^{k-1} \\ \ln \gamma_2 = \dfrac{x_1^2}{RT} \sum_{k=0}^{N} \left[1 - 2x_2 \left(1+k \right) \right] A_k \left(x_1 - x_2 \right)^{k-1} \end{cases} \end{cases} \qquad [3.30]$$

– Extension of the Redlich–Kister model to multicomponent systems containing more than two compounds:

We proceed in exactly the same way as for the Margules models. We combine equation [3.26] with:

$$g_{ij}^E = x_i x_j \sum_{k=0}^{N} A_{ij,k} \left(x_i - x_j \right)^k \qquad [3.31]$$

3.5.6. *The Van Laar model*

Van Laar (1860–1938) was a Dutch physicist–chemist, a student of Van der Waals. It is therefore entirely logical that the activity-coefficient model that he proposed is based on the use of the Van der Waals equation of state.

The assumptions that he formulated are:

– incompressible liquid (g^E does not depend on pressure). This is equivalent to neglecting the excess molar volume: $v^E = \left(\partial g^E / \partial P \right)_{T,x} = 0$;

– regular solution ($s^E = 0$: no combinatorial term). This being the case, g^E becomes equivalent to an excess enthalpy, itself equal to an excess internal energy:

$$g^E = h^E - Ts^E = u^E + P\underbrace{v^E}_{0} - T\underbrace{s^E}_{0} = u^E. \qquad [3.32]$$

The Van Laar model is therefore an enthalpic model.

The expression for u^E comes from the Van der Waals equation of state.

For a binary system, this leads to:

$$g^E_{\text{Van Laar}} = u^E = \frac{x_1 x_2 b_1 b_2}{x_1 b_1 + x_2 b_2} \left(\frac{\sqrt{a_1}}{b_1} - \frac{\sqrt{a_2}}{b_2} \right)^2 \qquad [3.33]$$

We set:

$$\begin{cases} A_{12} = b_1 \left(\dfrac{\sqrt{a_1}}{b_1} - \dfrac{\sqrt{a_2}}{b_2} \right)^2 \\[4mm] A_{21} = b_2 \left(\dfrac{\sqrt{a_1}}{b_1} - \dfrac{\sqrt{a_2}}{b_2} \right)^2 \end{cases} \qquad [3.34]$$

We therefore obtain:

$$g^E = u^E = x_1 \cdot x_2 \cdot \frac{A_{12} \cdot A_{21}}{A_{12} \cdot x_1 + A_{21} \cdot x_2} \qquad [3.35]$$

In practice, the two parameters A_{12} and A_{21} can be fitted to experimental data. The expressions for the activity coefficients are:

$$\begin{cases} \ln \gamma_1 = \dfrac{A_{12}}{RT} \left(\dfrac{A_{21} \cdot x_2}{A_{12} \cdot x_1 + A_{21} \cdot x_2} \right)^2 \\[4mm] \ln \gamma_2 = \dfrac{A_{21}}{RT} \left(\dfrac{A_{12} \cdot x_1}{A_{12} \cdot x_1 + A_{21} \cdot x_2} \right)^2 \end{cases} \qquad [3.36]$$

As previously seen, an alternative expression can be considered for the Van Laar model:

$$\frac{g^E}{RT} = x_1 \cdot x_2 \cdot \frac{A_{12} \cdot A_{21}}{A_{12} \cdot x_1 + A_{21} \cdot x_2} \quad and \quad \begin{cases} \ln \gamma_1 = A_{12} \left(\dfrac{A_{21} \cdot x_2}{A_{12} \cdot x_1 + A_{21} \cdot x_2} \right)^2 \\[4mm] \ln \gamma_2 = A_{21} \left(\dfrac{A_{12} \cdot x_1}{A_{12} \cdot x_1 + A_{21} \cdot x_2} \right)^2 \end{cases}$$

[3.37]

3.5.7. The Scatchard–Hildebrand (SH) model

This model originated approximately in the 1930s. The assumptions are the same as for the Van Laar model:

– incompressible liquid (g^E does not depend on the pressure), in other words $v^E = 0$;

– regular solution ($s^E = 0$). Consequently: $g^E = h^E = u^E$. **The SH model is an enthalpic model**.

The expression for u^E proposed by the authors for a binary system is:

$$g_{SH}^E = u_{SH}^E = (x_1 v_1 + x_2 v_2) \phi_1 \phi_2 (\delta_1 - \delta_2)^2 \quad with: \begin{cases} \phi_1 = \dfrac{x_1 v_1}{x_1 v_1 + x_2 v_2} \\[4mm] \phi_2 = \dfrac{x_2 v_2}{x_1 v_1 + x_2 v_2} \end{cases}$$

[3.38]

δ_i is known as the **Hildebrand solubility parameter** of the component i and is expressed in $J^{1/2} \cdot m^{-3/2}$. $\phi_i = x_i v_i / v$ is known as the **volume fraction** of the component i (dimensionless).

We should note that by setting out $\begin{cases} A_{12} = v_1 (\delta_1 - \delta_2)^2 \\ A_{21} = v_2 (\delta_1 - \delta_2)^2 \end{cases}$, we find that we have returned to the Van Laar model. The expressions for the activity coefficients of SH are:

$$\begin{cases} \ln \gamma_1 = \dfrac{v_1}{R \cdot T} \phi_2^2 \left(\delta_1 - \delta_2 \right)^2 \\ \\ \ln \gamma_2 = \dfrac{v_2}{R \cdot T} \phi_1^2 \left(\delta_1 - \delta_2 \right)^2 \end{cases}$$

[3.39]

3.5.8. *The NRTL (non-random two-liquid) model with three adjustable parameters per binary system*

This model was published in 1968. As for the two previous models, the authors of the NRTL model presumed the liquid phases to be incompressible (therefore $v^E = 0$) and proposed an expression for the excess internal energy.

We write g_{ij}, the energy of interaction between the molecules i and j such that $g_{ij} = g_{ji}$ and $g_{ii} = g_{jj} = 0$. For a binary system, the molar excess Gibbs energy is given by:

$$\frac{g^E}{RT}(T, x_1) = \frac{u^E}{RT}(T, x_1) = x_1 x_2 \left(\frac{\tau_{21} G_{21}}{x_1 + x_2 G_{21}} + \frac{\tau_{12} G_{12}}{x_2 + x_1 G_{12}} \right)$$

with
$$\begin{cases} \tau_{12} = \overbrace{\dfrac{g_{12} - g_{22}}{RT}}^{\text{denoted } b_{12}} \text{ and } G_{12} = \exp\left(-\alpha_{12} \tau_{12} \right) \\ \\ \tau_{21} = \overbrace{\dfrac{g_{21} - g_{11}}{RT}}^{\text{denoted } b_{21}} \text{ and } G_{21} = \exp\left(-\alpha_{12} \tau_{21} \right) \end{cases}$$

[3.40]

We should note that if we choose identical interaction energies $g_{11} = g_{22} = g_{12} = g_{21}$, we have $g^E = 0$. Consequently, NRTL asserts that the excess Gibbs energy of a system made up of molecules that have different sizes and shapes but for which the interactions 1–1, 2–2 and 1–2 are identical is zero. **We can thus conclude that the NRTL model does not take into account the size and shape effects.** The authors of the model also write: "the three-parameter NRTL model is more properly applicable to excess enthalpy than to excess Gibbs energy". With this phrase, they confirm that the NRTL model is an enthalpic model (or equivalently, an excess internal energy model):

$$g^E = \underbrace{g_{res}^E}_{\text{residual term}} = h^E = u^E + P\underbrace{v^E}_{0} = u^E \qquad [3.41]$$

The activity coefficients of the model are given by:

$$\begin{cases} \ln \gamma_1 = x_2^2 \left[\tau_{21} \left(\dfrac{G_{21}}{x_1 + x_2 G_{21}} \right)^2 + \dfrac{\tau_{12} G_{12}}{\left(x_2 + x_1 G_{12} \right)^2} \right] \\[4mm] \ln \gamma_2 = x_1^2 \left[\tau_{12} \left(\dfrac{G_{12}}{x_2 + x_1 G_{12}} \right)^2 + \dfrac{\tau_{21} G_{21}}{\left(x_1 + x_2 G_{21} \right)^2} \right] \end{cases} \qquad [3.42]$$

When applied to a binary system, the NRTL model contains three adjustable parameters: b_{12}, b_{21} and α_{12}.

In a multicomponent mixture (with p compounds), the NRTL model can be written as:

$$\frac{g^E}{R \cdot T} = \sum_{i=1}^{p} x_i \frac{\displaystyle\sum_{j=1}^{p} \tau_{ji} G_{ji} x_j}{\displaystyle\sum_{\ell=1}^{p} G_{\ell i} x_\ell} \quad \text{with} \quad \begin{cases} \tau_{ji} = \dfrac{g_{ji} - g_{ii}}{R \cdot T} = \dfrac{b_{ji}}{R \cdot T} \\[3mm] G_{ji} = \exp\left(-\alpha_{ji} \tau_{ji} \right) \text{ with } \alpha_{ji} = \alpha_{ij} \end{cases} \qquad [3.43]$$

The activity coefficients are written as:

$$\ln \gamma_i = \frac{\displaystyle\sum_{j=1}^{p} \tau_{ji} G_{ji} x_j}{\displaystyle\sum_{\ell=1}^{p} G_{\ell i} x_\ell} + \sum_{j=1}^{p} \frac{G_{ij} x_j}{\displaystyle\sum_{\ell=1}^{p} G_{\ell i} x_\ell} \left(\tau_{ij} - \frac{\displaystyle\sum_{r=1}^{m} \tau_{rj} G_{rj} x_r}{\displaystyle\sum_{\ell=1}^{p} G_{\ell i} x_\ell} \right) \qquad [3.44]$$

The NRTL model for multicomponent systems presents three adjustable parameters per binary system: b_{ij}, b_{ji} and $\alpha_{ij} = \alpha_{ji}$.

3.5.9. *NRTL models with four and six adjustable parameters per binary system*

For the model with four adjustable parameters per binary system, we set down $\alpha_{12} = \alpha_{21} = 0,2$ and we express the coefficients b_{12} and b_{21} according to:

$$\begin{cases} b_{12} = b_{12}^0 + b_{12}^1(T - T_0) \\ b_{21} = b_{21}^0 + b_{21}^1(T - T_0) \end{cases} \qquad [3.45]$$

T_0 is a reference temperature set arbitrarily (at 298 K, for example). The four adjustable parameters are then b_{12}^0, b_{12}^1, b_{21}^0, b_{21}^1.

For the model with six adjustable parameters per binary system, we write:

$$\begin{cases} b_{12} = b_{12}^0 + b_{12}^1(T - T_0) \\ b_{21} = b_{21}^0 + b_{21}^1(T - T_0) \\ \alpha_{12} = \alpha_{21} = \alpha_{21}^0 + \alpha_{21}^1(T - T_0) \end{cases} \qquad [3.46]$$

The six adjustable parameters are then b_{12}^0, b_{12}^1, b_{21}^0, b_{21}^1, α_{21}^0 and α_{21}^1.

3.5.10. *Wilson and Flory–Huggins models*

The Wilson model dates back to 1964. The liquid phases are presumed to be incompressible $(v^E = 0)$. It follows: $g^E = h^E - Ts^E = u^E + P\underset{0}{\underbrace{v^E}} - Ts^E = a^E$. The Wilson model can therefore be seen as a model of excess Helmholtz energy.

For a binary system, its expression is:

$$\frac{g^E}{R \cdot T} = -x_1 \cdot \ln\left(x_1 + x_2 \Lambda_{12}\right) - x_2 \cdot \ln\left(x_2 + x_1 \Lambda_{21}\right)$$

$$\text{with} \quad \begin{cases} \Lambda_{12} = \dfrac{v_2^*}{v_1^*} \exp\left(-\dfrac{\lambda_{12} - \lambda_{11}}{R \cdot T}\right) \\[4mm] \Lambda_{21} = \dfrac{v_1^*}{v_2^*} \exp\left(-\dfrac{\lambda_{21} - \lambda_{22}}{R \cdot T}\right) \end{cases} \qquad [3.47]$$

The quantities λ_{ij} designate the energies of interaction between the molecules i and j. We generally define two adjustable parameters:

$$\begin{cases} A_{12} = \lambda_{12} - \lambda_{11} \\ A_{21} = \lambda_{21} - \lambda_{22} \end{cases}.$$

Contrary to the Van Laar, Scatchard–Hildebrand and NRTL models that are purely enthalpic models, the Wilson model also contains a combinatorial term (size and shape term). To identify this term, we simply need to cancel out the enthalpic contribution by imposing: $\lambda_{12} = \lambda_{11} = \lambda_{22} \Rightarrow \begin{cases} A_{12} = 0 \\ A_{21} = 0 \end{cases}.$

In doing so, equation [3.47] becomes:

$$\begin{aligned} \frac{g^E_{comb}}{R \cdot T} &= -x_1 \cdot \ln\left(x_1 + x_2 \frac{v_2^*}{v_1^*}\right) - x_2 \cdot \ln\left(x_2 + x_1 \frac{v_1^*}{v_2^*}\right) \\[2mm] &= -x_1 \cdot \ln\left(\frac{x_1 v_1^* + x_2 v_2^*}{v_1^*}\right) - x_2 \cdot \ln\left(\frac{x_2 v_2^* + x_1 v_1^*}{v_2^*}\right) \qquad [3.48] \\[2mm] &= x_1 \cdot \ln\left(\frac{v_1^*}{v}\right) + x_2 \cdot \ln\left(\frac{v_2^*}{v}\right) \end{aligned}$$

We will now introduce the volume fractions defined by equation [3.38] into [3.48].

$$\frac{g^E_{comb}}{R \cdot T} = x_1 \cdot \ln\left(\frac{\phi_1}{x_1}\right) + x_2 \cdot \ln\left(\frac{\phi_2}{x_2}\right) \quad \text{(always a negative quantity)} \qquad [3.49]$$

This expression [3.49] is the expression for the **Flory–Huggins activity-coefficient model**. The expression for the enthalpic (or residual) contribution of the Wilson model can then be obtained simply by:

$$\frac{g_{\text{rés}}^E}{R \cdot T} = \frac{g^E}{R \cdot T} - \frac{g_{\text{comb}}^E}{R \cdot T}$$

$$= -x_1 \cdot \ln\left((x_1 + x_2 \Lambda_{12}) \frac{\phi_1}{x_1} \right) - x_2 \cdot \ln\left((x_2 + x_1 \Lambda_{21}) \frac{\phi_2}{x_2} \right)$$

[3.50]

In the most general case of a mixture of p components, the g^E expression for the Wilson model is written as:

$$\left\{ \begin{array}{l} \dfrac{g^E}{R \cdot T} = -\displaystyle\sum_{i=1}^{p} x_i \ln\left(\sum_{j=1}^{p} x_j \Lambda_{ij} \right) \\[3mm] \ln \gamma_i = -\ln\left(\displaystyle\sum_{j=1}^{p} x_j \Lambda_{ij} \right) + 1 - \displaystyle\sum_{k=1}^{p} \dfrac{x_k \Lambda_{ki}}{\displaystyle\sum_{j=1}^{p} x_j \Lambda_{kj}} \\[5mm] \text{with } \Lambda_{ij} = \dfrac{v_j^*}{v_i^*} \exp\left(-\dfrac{\lambda_{ij} - \lambda_{ii}}{R \cdot T} \right) \end{array} \right.$$

[3.51]

As previously mentioned, there are two adjustable parameters for each binary system in the mixture: $\left\{ \begin{array}{l} A_{ij} = \lambda_{ij} - \lambda_{ii} \\ A_{ji} = \lambda_{ji} - \lambda_{jj} \end{array} \right.$.

We mention at this point a significant characteristic of the Wilson model: **this model is excellent for representing VLEs of mixtures but does not have the capacity to predict liquid–liquid equilibria.** This is a limitation that must always be kept in mind when selecting an activity-coefficient model for a given application.

3.5.11. *UNIQUAC model*

The UNIQUAC model was proposed in 1975. As with all activity-coefficient models, the assumption is made that the liquids are incompressible: $v^E = \left(\partial g^E / \partial P \right)_{T,x} = 0$. As for the Wilson model, we therefore have $g^E = a^E$.

In the case of a binary system, it is written as:

$$\begin{cases} \dfrac{g^E}{RT} = \overbrace{x_1 \ln \dfrac{\varphi_1}{x_1} + x_2 \ln \dfrac{\varphi_2}{x_2}}^{\text{Flory-Huggins}} + \overbrace{\dfrac{z}{2}\left(q_1 x_1 \ln \dfrac{\theta_1}{\varphi_1} + q_2 x_2 \ln \dfrac{\theta_2}{\varphi_2} \right)}^{\text{Staverman-Guggenheim}} \\[2pt] \underbrace{-q_1 x_1 \ln\left(\theta_1 + \theta_2 \tau_{21}\right) - q_2 x_2 \ln\left(\theta_2 + \theta_1 \tau_{12}\right)}_{g^E_{res}/(RT)} \\[2pt] \tau_{12} = \exp\left(-\dfrac{u_{12} - u_{22}}{R \cdot T}\right) \qquad \tau_{21} = \exp\left(-\dfrac{u_{21} - u_{11}}{R \cdot T}\right) \\[2pt] q_i = \text{surface area of molecule } i \qquad r_i = \text{volume of molecule } i \\[2pt] \theta_1 = \dfrac{q_1 x_1}{q_1 x_1 + q_2 x_2} \qquad \theta_2 = \dfrac{q_2 x_2}{q_2 x_2 + q_2 x_2} \qquad \text{(area fractions)} \\[2pt] \varphi_1 = \dfrac{r_1 x_1}{r_1 x_1 + r_2 x_2} \qquad \varphi_2 = \dfrac{r_2 x_2}{r_2 x_2 + r_2 x_2} \qquad \text{(volume fractions)} \end{cases}$$

$$\overbrace{g^E_{comb}/(RT)}$$

[3.52]

The values r_i and q_i that intervene in this relationship are (adimensional) measurements of the volume and surface area of the molecule i. They correspond to the Van der Waals volume and surface area, defined by Bondi *et al.* divided respectively by a fixed volume and reference surface area of $15.17 \text{ cm}^3 \cdot \text{mol}^{-1}$ and $2.5 \cdot 10^9 \text{ cm}^2 \cdot \text{mol}^{-1}$. For more details, refer to the original publication of the model[3]. Some values are provided in Table 3.1.

Compounds i	r_i	q_i
CCl$_4$	3.33	2.82
CH$_3$–OH	1.43	1.43
H$_3$C–CN	1.87	1.72
C$_2$H$_6$–NO$_2$	2.68	2.41
H$_3$C–CO–CH$_3$	2.57	2.34
Benzene	3.19	2.40
Methyl cyclopentane	3.87	3.01
n-heptane	5.17	4.40

Table 3.1. *Values of coefficients r and q in the UNIQUAC model for some molecules*

3 D.S. Abrams, J.M. Prausnitz, *AIChE Journal*, 21, 1, 116–128 (1975).

The UNIQUAC model defines two adjustable parameters per binary system: $\begin{cases} b_{12} = u_{12} - u_{22} \\ b_{21} = u_{21} - u_{11} \end{cases}$, where u_{ij} defines an energy of interaction between molecules i and j. As expected, in the case where the molecular interactions all become identical ($u_{12} = u_{22} = u_{21} = u_{11}$), the residual (enthalpic) contribution disappears and the molar excess Gibbs energy of the binary system is thus associated exactly with the combinatorial part of the model: $\begin{cases} g_{res}^E = 0 \\ g^E = g_{comb}^E \end{cases}$.

The activity coefficients are given by:

$$
\begin{cases}
\ln \gamma_1 = \underbrace{\ln \frac{\varphi_1}{x_1} + \frac{z}{2} q_1 \ln \frac{\theta_1}{\varphi_1} + \varphi_2 \left(L_1 - \frac{r_1}{r_2} L_2 \right)}_{\ln \gamma_1^{comb}} \\
\qquad \underbrace{-q_1 \ln\left(\theta_1 + \theta_2 \tau_{21}\right) + \theta_2 q_1 \left(\frac{\tau_{21}}{\theta_1 + \theta_2 \tau_{21}} - \frac{\tau_{12}}{\theta_1 \tau_{12} + \theta_2} \right)}_{\ln \gamma_1^{res}} \\[2em]
\ln \gamma_2 = \underbrace{\ln \frac{\varphi_2}{x_2} + \frac{z}{2} q_2 \ln \frac{\theta_2}{\varphi_2} + \varphi_1 \left(L_2 - \frac{r_2}{r_1} L_1 \right)}_{\ln \gamma_2^{comb}} \\
\qquad \underbrace{-q_2 \ln\left(\theta_2 + \theta_1 \tau_{12}\right) + \theta_1 q_2 \left(\frac{\tau_{12}}{\theta_1 \tau_{12} + \theta_2} - \frac{\tau_{21}}{\theta_1 + \theta_2 \tau_{21}} \right)}_{\ln \gamma_2^{res}}
\end{cases}
\qquad [3.53]
$$

In the case of a multicomponent mixture (with p compounds), the UNIQUAC model adopts the following expression:

$$\begin{cases} g^E = g^E_{comb} + g^E_{res} \\[2mm] \dfrac{g^E_{comb}}{R \cdot T} = \sum_{i=1}^{p} x_i \ln\left(\dfrac{\varphi_i}{x_i}\right) + \dfrac{z}{2} \sum_{i=1}^{p} q_i x_i \ln\left(\dfrac{\theta_i}{\varphi_i}\right) \\[2mm] \dfrac{g^E_{res}}{R \cdot T} = -\sum_{i=1}^{p} q_i x_i \ln\left(\sum_{i=1}^{p} \theta_j \tau_{ji}\right) \\[2mm] \tau_{ji} = \exp\left[-\left(u_{ji} - u_{ii}\right)/(RT)\right] \\[2mm] \theta_i = q_i x_i \Big/ \sum_j q_j x_j \\[2mm] \varphi_i = r_i x_i \Big/ \sum_j r_j x_j \end{cases}$$ [3.54]

The activity coefficients are given by:

$$\begin{cases} \ln \gamma_i = \ln \gamma_i^{comb} + \ln \gamma_i^{res} \\[2mm] \ln \gamma_i^{comb} = \ln\left(\dfrac{\varphi_i}{x_i}\right) + \dfrac{z}{2} \cdot q_i \cdot \ln\left(\dfrac{\theta_i}{\varphi_i}\right) + L_i - \dfrac{\varphi_i}{x_i} \sum_{j=1}^{p} x_j L_j \\[2mm] \ln \gamma_i^{res} = q_i \left[1 - \ln\left(\sum_{i=1}^{p} \theta_j \tau_{ji}\right) - \sum_{j=1}^{p} \dfrac{\theta_j \tau_{ij}}{\sum\limits_{k=1}^{p} \theta_k \tau_{kj}} \right] \\[2mm] L_i = \dfrac{z}{2}\left(r_i - q_i\right) - \left(r_i - 1\right) \end{cases}$$ [3.55]

In the expressions [3.54] and [3.55], the two adjustable parameters per binary system are: $\begin{cases} b_{ij} = u_{ij} - u_{jj} \\ b_{ji} = u_{ji} - u_{ii} \end{cases}$

3.5.12. *The original UNIFAC model and its extensions*

This model dating back to 1975 is a **predictive version** of the UNIQUAC model, meaning that this model does not require knowledge of the binary interaction parameters b_{ij} and b_{ji} found in the UNIQUAC model. The only input parameters for the model are the chemical structures of the molecules in the mixture, in addition to the temperature and the composition. The chemical structure of each molecule is expressed by the concept of **group contributions** (as a reminder, this means that the various molecules are divided into functional groups and that the global properties of the solution are determined based on the properties of the groups). The molar excess Gibbs energy is typically the sum of two contributions, one combinatorial, the other residual:

$$\begin{cases} g^E_{UNIFAC} = g^E_{comb\,UNIQUAC} + g^E_{res,UNIFAC} \\ \ln \gamma_{i,UNIFAC} = \ln \gamma^{comb,UNIQUAC}_i + \ln \gamma^{res,UNIFAC}_i \end{cases} \qquad [3.56]$$

The combinatorial term is from the UNIQUAC model (see equation [3.54]). To estimate it, the adimensional surface areas and volumes of the molecules (q_i and r_i) must be known. They are estimated from group-contributions methods. The total number of groups defined by the method is denoted as NGS:

$$\begin{cases} r_i = \sum_{k=1}^{NGS} v_k^{(i)} R_k \\ q_i = \sum_{k=1}^{NGS} v_k^{(i)} Q_k \end{cases} \qquad [3.57]$$

where $v_k^{(i)}$ designates the occurrence of the group k in the molecule i (this is the number of times the group k appears in i). R_k and Q_k designate respectively the dimensionless volume and surface area *of the group k*. They are tabulated in the original article for the model[4].

---

4 A. Fredenslund et al. *AIChE Journal*, 21, 6, 1086–1099 (1975).

The residual term is estimated using equation [3.58]:

$$
\begin{cases}
g_{res,UNIFAC}^{E} = RT \sum_{i} x_i \ln \ln \gamma_i^{res,UNIFAC} \\
\ln \gamma_i^{res,UNIFAC} = \sum_{k=1}^{NGS} v_k^{(i)} \left[\ln \Gamma_k - \ln \Gamma_k^{(i)} \right]
\end{cases}
\qquad [3.58]
$$

where Γ_k represents the activity coefficient *of the group* k in the liquid solution, whereas $\Gamma_k^{(i)}$ designates the activity coefficient *of the group* k in pure liquid i. The expressions for these quantities are:

$$
\begin{cases}
\ln \Gamma_k = Q_k \left[1 - \ln \left(\sum_{j=1}^{NGS} \theta_j \psi_{jk} \right) - \sum_{j=1}^{NGS} \frac{\theta_j \psi_{kj}}{\sum_{\ell=1}^{NGS} \theta_\ell \psi_{\ell j}} \right] \\
\ln \Gamma_k^{(i)} = Q_k \left[1 - \ln \left(\sum_{j=1}^{NGS} \theta_j^{(i)} \psi_{jk} \right) - \sum_{j=1}^{NGS} \frac{\theta_j^{(i)} \psi_{kj}}{\sum_{l=1}^{NGS} \theta_l^{(i)} \psi_{lj}} \right]
\end{cases}
\qquad [3.59]
$$

where:

θ_j represents the surface-area fraction of the group j in the mixture in question: $\theta_j = Q_j X_j \Big/ \sum_{k=1}^{NGS} Q_k X_k$, with X_k, the molar fraction of the group k

in the mixture in question, given by: $X_k = \dfrac{\sum_{i=1}^{p} x_i v_k^{(i)}}{\sum_{i=1}^{p} x_i \left(\sum_{j=1}^{NGS} v_j^{(i)} \right)}$.

The parameter ψ_{jk} characterizes the interaction of the groups j and k in solution: $\psi_{jk} = \exp\left(-\dfrac{a_{m_j n_k}}{T} \right)$. The values of $a_{m_j n_k}$ are tabulated in the original article by the authors of the model.

$\theta_j^{(i)}$ designates the surface fraction of the group j in a solution of pure i:

$$\theta_j^{(i)} = Q_j X_j^{(i)} \Big/ \sum_{k=1}^{NGS} Q_k X_k^{(i)} \text{ and } X_k^{(i)} \text{ is the molar fraction of the group } k \text{ in the}$$

mixture in question: $X_k^{(i)} = \dfrac{v_k^{(i)}}{\displaystyle\sum_{j=1}^{NGS} v_j^{(i)}}$.

Currently, the UNIFAC model remains a particular favorite due to its predictive nature. Of course, better precision than that produced by UNIFAC is achieved when a model whose parameters have been fitted to well-chosen experimental data is used. However, if no experimental data are available, the UNIFAC model has the capacity to provide reasonable estimations of the thermodynamic properties and in particular, of the behaviors of phases; from a quantitative point of view, we generally consider that the errors induced by the use of this model are approximately 5–10%.

Many updated versions of UNIFAC have been proposed since it was created. The most developed version is certainly UNIFAC-Dortmund[5].

3.5.13. *Summary: an aid to selecting an activity-coefficient model*

This is a brief summary of the main points in this chapter and, in particular, of the paragraph regarding activity-coefficient models.

– Concerning the VLE relationship at low pressure:

The VLE relationship [3.4] can only be applied if the saturated-vapor pressures of the two components can be defined. In the case of a binary system, this involves $T < \min\{T_{c,1}; T_{c,2}\}$ and $P < \min\{P_{c,1}; P_{c,2}\}$.

---

5 U. Weidlich, J. Gmehling, *Ind. Eng. Chem. Res.* 26, 1373 (1987).

The VLE relationship [3.4] uses two different models for the liquid (activity-coefficient model also known as "g^E model") and the gas (equation of state) phases. This approach does not account for continuity between the liquid and gas states and therefore can only be applied in the absence of a vapor–liquid critical point. This again implies for a binary system that $\begin{cases} T < \min\{T_{c,1}\,;T_{c,2}\} \\ P < \min\{P_{c,1}\,;P_{c,2}\} \end{cases}$. As explained at the beginning of this chapter, it is necessary to apply a security margin to these expressions of inequality because binary critical points can exist at pressures that are slightly lower than $\min\{P_{c,1}\,;P_{c,2}\}$ and at temperatures that are slightly lower than $\min\{T_{c,1}\,;T_{c,2}\}$.

In 95% of cases, we presume that the coefficient C_i of the relationship [3.4] is equal to 1 (reasonable hypothesis at moderate pressure, i.e. when $P < 20\,\text{bar}$).

At low temperature and low pressure, applying the hypotheses described above, the VLE relationship is written as $P \cdot y_i = P_i^{sat}(T) \cdot x_i \cdot \gamma_i(T,P,x)$. The term $P_i^{sat}(T)$ can be estimated using a correlation (Antoine, Clapeyron, etc.), whereas the activity coefficient γ_i requires the use of an activity-coefficient model. If the liquid solution in question is ideal (molecules of similar sizes, shapes and interactions), then $\gamma_i = 1$.

– Concerning activity-coefficient models:

Activity-coefficient models have the ability to represent systems that are weakly or highly polar, whether associating (by hydrogen bonds) or not.

All activity-coefficient models that are not predictive require knowledge of **binary interaction parameters**, present in the residual part. In the context of process simulation, any binary interaction parameter that is not specified means that the corresponding binary mixture is presumed athermic ($h^E = 0$) or even ideal for purely enthalpic models (without a combinatorial part).

– **Classification of activity-coefficient models:**

Purely correlative models: Margules (mainly academic interest), Redlich–Kister (usable but it is advised to give preference to a model with a physical basis for a more reliable extrapolation in temperature)

Models with a physical basis that use adjustable parameters:

– purely enthalpic models (interactions but no size effects): Scatchard–Hildebrand, **NRTL**;

– purely entropic models (no interaction; size effects): **Flory–Huggins** (ideal for systems that contain a monomer and its polymer);

– mixed models: **Wilson** (although it cannot represent LLEs), **UNIQUAC**.

Predictive models (to be used as a last resort, in total absence of experimental information): **UNIFAC** (group-contribution method).

3.6. Calculation of the vapor–liquid equilibrium of binary systems at low temperature and low pressure

This paragraph is a short presentation of the calculation methods for VLE isothermal and isobaric phase diagrams of binary systems (keeping things simple, we will not give details of the calculation of LLE or VLLE). Two approaches are mainly used in practice:

A basic VLE calculation (bubble-point curve or dew-point curve): as previously discussed, the VLE condition is expressed in two equations:

$$\begin{cases} P \cdot y_1 = P_1^{sat}(T) \cdot x_1 \cdot \gamma_1(T,P,\boldsymbol{x}) \\ P \cdot (1 - y_1) = P_2^{sat}(T) \cdot (1 - x_1) \cdot \gamma_2(T,P,\boldsymbol{x}) \end{cases} \qquad [3.60]$$

These two equations involve four variables: T, P, x_1 and y_1. The variance of the system, which is its degree of freedom, is therefore equal to 2; in other words, two variables among the four must be specified so that the problem can be resolved. The main combinations of specified variables are listed in Table 3.2 and will be explained in more detail in the following paragraphs.

We specify	We calculate	Name of calculation
T and x_1	P and y_1	**Bubble-point pressure** (**well suited for calculation of a** *Pxy* **diagram**)
T and y_1	P and x_1	Dew-point pressure
P and x_1	T and y_1	**Bubble-point temperature (well suited for calculation of a** *Txy* **diagram)**
P and y_1	T and x_1	Dew-point temperature

Table 3.2. *Four main types of calculation of VLE binary systems*

PT-flash calculation:

In equation [3.60], only the intensive variables of the phases are present. In the case of a PT-flash calculation, in addition to the temperature and pressure, the overall composition of the system (z_1) has to be specified. The objective of the flash calculation is to determine the physical state of the system (single-phase or two-phase) and if it is two-phase to determine the composition and proportion of the phases that are in equilibrium. The equations to be solved are the two equations labeled [3.60] and a *material balance*:

$$z_1 = \tau y_1 + (1-\tau) x_1$$

$$\begin{cases} z_1 \text{ designates the global molar fraction of constituent 1,} \\ \text{and } \tau = \dfrac{n_{gaz}}{n_{liq} + n_{gaz}} \text{ the molar proportion of gas phase.} \end{cases} \qquad [3.61]$$

The system made up of equations [3.60] and [3.61] then involves six variables: T, P, x_1, y_1, z_1 and τ. Among these variables, three are specified (T, P, z_1) and three (x_1, y_1, τ) are calculated by solving the three equations of the system.

We will now describe the two types of approach.

3.6.1. *Bubble-point pressure calculation*

Equations to solve: [3.60]. We specify T and x_1; we calculate P and y_1.

Method of resolution:

We separate the problem into steps. We begin by adding together the two equations to be solved in order to get an explicit expression for the VLE pressure:

$$\begin{cases} P_{VLE} \cdot y_1 = P_1^{sat}(T) \cdot x_1 \cdot \gamma_1(T, x_1) \\ P_{VLE} \cdot (1 - y_1) = P_2^{sat}(T) \cdot (1 - x_1) \cdot \gamma_2(T, x_1) \end{cases} \qquad [3.62]$$

$$\Sigma : P_{VLE} = P_1^{sat}(T) \cdot x_1 \cdot \gamma_1(T, x_1) + P_2^{sat}(T) \cdot (1 - x_1) \cdot \gamma_2(T, x_1)$$

We then calculate y_1 from one of the two equations of the system [3.60]; for example:

$$y_1 = \frac{P_1^{sat}(T) \cdot x_1 \cdot \gamma_1(T, x_1)}{P_{VLE}} \qquad [3.63]$$

The result of the VLE calculation is illustrated in Figure 3.32. This type of calculation is particularly suitable for drawing a *Pxy* isothermal phase diagram using spreadsheet software. The drawing method is summarized by Figure 3.33. The bubble-point curve is obtained by plotting the vapor–liquid equilibrium pressure P_{VLE} as a function of x_1; the dew-point curve is obtained by plotting the same pressure P as a function of y_1.

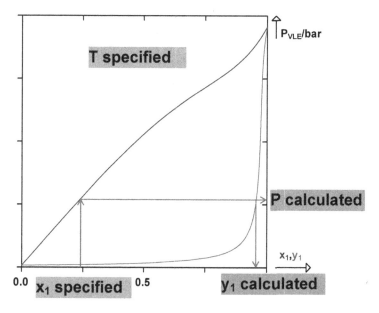

Figure 3.32. *Calculation principle of the bubble-point pressure. For a color version of the figure, see www.iste.co.uk/jaubert/thermodynamic.zip*

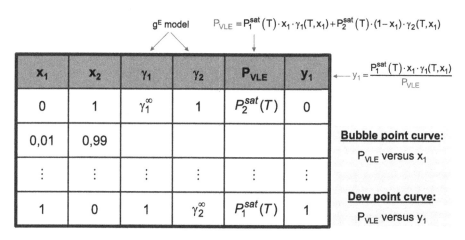

Figure 3.33. *Method of drawing a Pxy diagram using a spreadsheet and a bubble-pressure algorithm. For a color version of the figure, see www.iste.co.uk/jaubert/thermodynamic.zip*

NOTE.– When the procedure to draw a *Pxy* diagram as explained above is applied to a system that presents liquid–liquid immiscibility, the isothermal diagram is characterized by the existence of metastable and unstable parts. This type of diagram can easily be recognized due to the presence of two cusps that mark the transition between the metastable and unstable domains. This is illustrated in Figure 3.34.

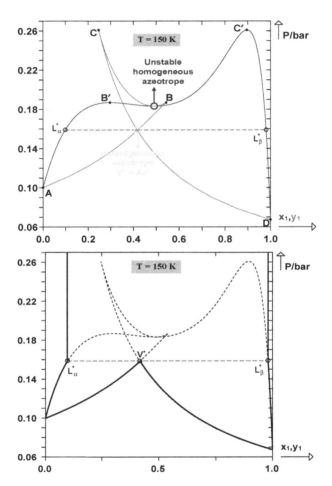

Figure 3.34. *Form of a VLE diagram with liquid–liquid demixing, drawn using a bubble-point pressure algorithm. Left: diagram containing stable, metastable and unstable sections of the curves. Right: only the stable sections of the curves have been maintained (solid lines); the unstable sections of the curves are indicated by the dotted lines. For a color version of the figure, see www.iste.co.uk/jaubert/ thermodynamic.zip*

3.6.2. *Calculation of the dew-point pressure*

Equations to solve: [3.60]. We specify T and y_1; we calculate P_{ELV} and x_1.

Method of resolution:

We separate the problem into steps. We begin by expressing P_{VLE} using the two equations in the system [3.60], then we eliminate this variable by subtracting the two equations. We then obtain a nonlinear equation with one unknown: x_1. This equation must be resolved by a numerical method (dichotomy, Newton, etc.):

$$
\begin{cases}
P_{VLE} = \dfrac{P_1^{sat}(T) \cdot x_1 \cdot \gamma_1(T,x_1)}{y_1} \\[4mm]
P_{VLE} = \dfrac{P_2^{sat}(T) \cdot (1-x_1) \cdot \gamma_2(T,x_1)}{1-y_1}
\end{cases}
$$

By subtraction: $0 = \dfrac{P_1^{sat}(T) \cdot x_1 \cdot \gamma_1(T,x_1)}{y_1} - \dfrac{P_2^{sat}(T) \cdot (1-x_1) \cdot \gamma_2(T,x_1)}{1-y_1}$

$$[3.64]$$

We then calculate P_{VLE} from one of the equations in system [3.60]; for example:

$$
P_{VLE} = \frac{P_1^{sat}(T) \cdot x_1 \cdot \gamma_1(T,x_1)}{y_1}
\qquad [3.65]
$$

3.6.3. *Calculation of the bubble-point temperature*

Equations to solve: [3.60]. We specify P and x_1; we calculate T_{VLE} and y_1.

Method of resolution:

We separate the problem into steps. We start with equation [3.62], with unknown T_{VLE} that we then resolve using a numerical method (Newton, dichotomy, etc.). We then calculate y_1 using equation [3.63].

The calculation principle is summarized in Figure 3.35. This type of calculation is particularly suitable to a *Txy* isobaric phase diagram drawn using spreadsheet software. The drawing method is similar to the one summarized in Figure 3.33 (*mutatis mutandis*).

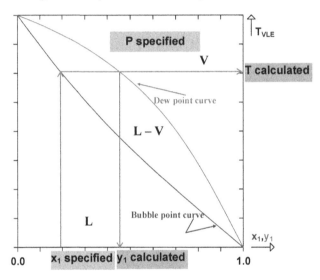

Figure 3.35. *Calculation principle of the bubble-point temperature. For a color version of the figure, see www.iste.co.uk/jaubert/thermodynamic.zip*

3.6.4. *Calculation of the dew-point temperature*

Equations to solve: [3.60]. We specify P and y_1; we calculate T_{VLE} and x_1.

Solution: there is no simple method that will allow a sequential resolution of the problem. The two equations in system [3.60] must be resolved simultaneously using a numerical method (Newton–Raphson, for example).

3.6.5. *PT-flash calculation*

Equations to solve:

$$\begin{cases} \text{VLE:} \begin{cases} P \cdot y_1 = P_1^{sat}(T) \cdot x_1 \cdot \gamma_1(T,P,x_1) \\ P \cdot (1-y_1) = P_2^{sat}(T) \cdot (1-x_1) \cdot \gamma_2(T,P,x_1) \end{cases} \\ \text{Material balance: } z_1 = \tau y_1 + (1-\tau)x_1 \Leftrightarrow \tau = \dfrac{z_1 - x_1}{y_1 - x_1} \end{cases} \qquad [3.66]$$

We specify T, P and z_1; we calculate x_1, y_1 and τ. The principle of the PT-flash method is summarized in Figure 3.36.

Solution: there are many methods used to solve the PT-flash problem. The most well known is certainly the Rachford–Rice method. To remain concise, we will not go into details of the procedure.

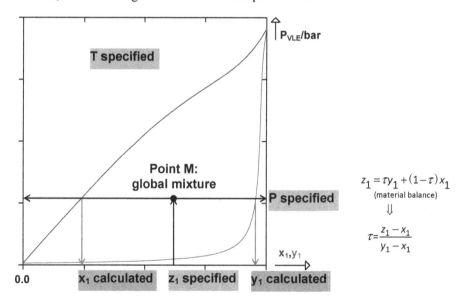

Figure 3.36. *Principle of a PT-flash calculation. For a color version of the figure, see www.iste.co.uk/jaubert/thermodynamic.zip*

Estimation of the Thermodynamic Properties of Mixtures from an Equation of State: An Overview of Models and Calculation Procedures

4.1. The phase-equilibrium condition and the φ-φ approach

The equilibrium condition between uniform phases of a multicomponent system in VLE was given in the previous chapter (see equation [3.2]). Remembering that the temperature and pressure are the same in all phases of the system, this condition is written simply as:

$$\mu_{i,liq}(T,P,\mathbf{x}) = \mu_{i,vap}(T,P,\mathbf{y}) \qquad \forall \ component \ i \qquad [4.1]$$

where μ_i denotes the chemical potential of component i in the mixture. By definition, the fugacity f_i of a component i in a uniform phase at $(T,P,\mathbf{z})$ – where $\mathbf{z}$ denotes the vector of molar fractions of the components – is given by the equation:

$$\underbrace{\mu_i(T,P,\mathbf{z})}_{\substack{\text{chemical potential} \\ \text{of component } i \\ \text{in the mixture}}} = \underbrace{g^{\bullet}_{i\,\text{pur}}(T,P)}_{\substack{\text{chemical potential of} \\ \text{PURE perfect gas } i}} + RT \ln\left[\frac{f_i(T,P,\mathbf{z})}{P}\right] \qquad [4.2]$$

By combining equations [4.1] and [4.2], we reach the phase-equilibrium condition:

$$\boxed{f_{i,liq}(T,P,\mathbf{x}) = f_{i,vap}(T,P,\mathbf{y})} \tag{4.3}$$

This condition [4.3] can also be expressed in terms of fugacity coefficients. Knowing that:

$$f_i(T,P,\mathbf{z}) = P \cdot z_i \cdot \varphi_i \tag{4.4}$$

the condition [4.3] can be rewritten:

$$\boxed{x_i \cdot \varphi_{i,liq}(T,P,\mathbf{x}) = y_i \cdot \varphi_{i,vap}(T,P,\mathbf{y})} \tag{4.5}$$

Fugacity coefficients φ_i (like fugacities f_i) are values that can be estimated from **equations of state for mixtures**. The approach, which consists of performing a vapor–liquid equilibrium calculation (i.e. calculating the fugacity coefficients of the components in these phases) using one single equation of state, is known as the "*phi–phi approach (φ-φ)*".

4.2. General presentation of the usual volumetric equations of state applicable to mixtures

For a mixture, a volumetric equation of state is a relationship of the form:

$$f(P,v,T,\mathbf{z}) = 0 \tag{4.6}$$

i.e. a mathematical relationship between pressure, total molar volume, temperature and molar fractions.

In the same way as for pure substances, there are two main categories of equations of state for mixtures:

Volume-explicit equations of state: these are of the form $v = f(T,P,\mathbf{z})$. *These equations can only be used to describe gas phases.* We will mention, for example, the truncated virial equation of state.

Pressure-explicit equations of state: these are of the form $P = f(T, v, \mathbf{z})$. These equations can be applied to gas and liquid phases and can represent VLE (among other things). Among the main pressure-explicit equations of state, we find:

– cubic equations of state;

– SAFT equations of state;

– equations of state that are specific to certain mixtures (Span–Wagner, GERG, etc.).

We will now discuss the differences and similarities between the equations of state for pure substances (mentioned in Chapter 2) and the equations of state for mixtures.

When they are extended to mixtures, equations of state maintain the same mathematical form as for a pure substance (*one-fluid theory*), but **mixing rules** must be defined in order to express the parameters in the equations of state for mixtures as a function of the parameters found in the equations of state for pure substances.

If the equation of state is volume-explicit, similar to, for example, the truncated virial equation of state, then it can be applied to a gaseous system and used in the γ-φ approach. It cannot be used in the φ-φ approach.

If the equation of state is pressure-explicit like cubic models or SAFT models, then it can be applied indifferently to a liquid or gaseous phase and can be used in the γ-φ and φ-φ approaches.

4.3. Virial expansions

In this section, we concentrate on the truncated virial equation, which is the most frequently used virial expansion in practice. Whether applied to a pure substance or a mixture, the truncated virial equation maintains the same mathematical form (by application of the *one-fluid theory*):

$$\begin{cases} \text{For a pure substance } i: \quad v_i^*(T, P) = RT/P + B_i(T) \\ \text{For a mixture:} \quad v(T, P, \mathbf{z}) = RT/P + B_m(T, \mathbf{z}) \end{cases} \qquad [4.7]$$

It is therefore necessary to define a *mixing rule* for the parameter B_m, in other words express it as a function of the second virial coefficients for pure substances, and of the composition. In our case here, statistical thermodynamics indicates that B_m must be expressed as a quadratic form of the molar fractions:

$$B_m(T,\mathbf{z}) = \sum_{i=1}^{p}\sum_{j=1}^{p} z_i \cdot z_j \cdot B_{ij}(T) \quad \text{with} \quad B_{ij}(T) = B_{ji}(T) \qquad [4.8]$$

where p denotes the number of components in the mixture, B_{ij} is known as the *second mutual virial coefficient* of components i and j (when $i \neq j$) and B_{ii} denotes the second virial coefficient of the pure substance i (this is the coefficient B_i of equation [4.7]). For a binary system, we would obtain:

$$v(T,P,y_1,y_2) = \frac{RT}{P} + B_{11} \cdot y_1^2 + B_{22} \cdot y_2^2 + 2B_{12}y_1y_2 \qquad [4.9]$$

4.4. Cubic equations of state

4.4.1. Generalities

The general expression of a cubic equation of state is written as:

$$
\begin{cases}
\text{For a pure substance } i: P(T,v) = \dfrac{RT}{v-b_i} - \dfrac{a_i(T)}{(v-b_ir_1)(v-b_ir_2)} \\[3mm]
\text{For a mixture:} \quad P(T,v,\mathbf{z}) = \dfrac{RT}{v-b_m(\mathbf{z})} - \dfrac{a_m(T,\mathbf{z})}{[v-b_m(\mathbf{z})\cdot r_1][v-b_m(\mathbf{z})\cdot r_2]}
\end{cases}
$$

$$[4.10]$$

In the same way as for the virial equation, *mixing rules* remain to be defined for the parameters a_m and b_m.

4.4.2. Classic mixing rules (known as "Van der Waals" mixing rules)

They have been constructed under the assumption that the mixture of molecules (1) and (2) of a binary system is random, which implies that in using a *molecule pin*, the probability of selecting a molecule (1) is equal to its molar fraction x_1 in the mixture. As an example, let us determine a mixing rule for the parameter a_m, which represents the attractive interactions. In all binary mixtures, there are three types of possible interactions between the two molecules: interactions (1)+(1), (2)+(2) or (1)+(2). The probability that a molecule (1) will interact with another molecule (1) is equal to x_1^2 because for independent events, the probability of an intersection is the product of the probabilities. Similarly, the probability of a (2)+(2) interaction is equal to x_2^2 and the probability of a (1)+(2) or (2)+(1) interaction is equal to $2x_1x_2$. If a_{11}, a_{22} and a_{12} represent respectively the contributions made by interactions (1)+(1), (2)+(2) and (1)+(2) to the interactions of the mixture, the mixing rule for a_m – which sets the mathematical method used to sum the contributions of the interactions – is written (for a random mixture) as:

$$a_m = x_1^2 a_{11} + 2x_1 x_2 a_{12} + x_2^2 a_{22} = \sum_{i=1}^{2}\sum_{j=1}^{2} x_i x_j a_{ij} \qquad [4.11]$$

where a_{11} and a_{22} are the attractive parameters of the pure substances (1) and (2) respectively, and here it is implied that a_{12} is identical to a_{21}.

Thus, classic mixing rules stipulate that the parameters a_m and b_m are quadratic functions of the composition:

$$\begin{cases} a_m = \sum_{i=1}^{p}\sum_{j=1}^{p} x_i x_j a_{ij} \\[2ex] b_m = \sum_{i=1}^{p}\sum_{j=1}^{p} x_i x_j b_{ij} \end{cases} \qquad [4.12]$$

The combining rules define the expressions for coefficients a_{ij} and b_{ij} of equation [4.12].

$$\begin{cases} a_{ij} = \sqrt{a_i a_j}\,(1 - k_{ij}) \\ b_{ij} = \tfrac{1}{2}\left(b_i + b_j\right)(1 - \ell_{ij}) \end{cases} \qquad [4.13]$$

These rules introduce two binary interaction parameters: k_{ij} and ℓ_{ij}; k_{ij} is by far the most important, whereas we often set $\ell_{ij} = 0$. In this particular case:

$$b_m = \sum_{i=1}^{p} \sum_{j=1}^{p} x_i x_j \times \tfrac{1}{2}\left(b_i + b_j\right) = \sum_{i=1}^{p} x_i b_i \qquad [4.14]$$

In order to illustrate the importance of the parameter k_{ij}, we will look at the system 2,2,4 trimethylpentane (1) + toluene (2) at 333 K, represented by the Peng–Robinson (PR) equation of state armed with classic mixing rules. Figure 4.1 highlights the influence of k_{ij} on phase behavior. In this example, we should note in particular that $k_{12} = 0$ does not imply an ideal-liquid solution (although by canceling the binary interaction parameters of purely enthalpic activity-coefficient models, we systematically obtain an *ideal-solution* type of behavior).

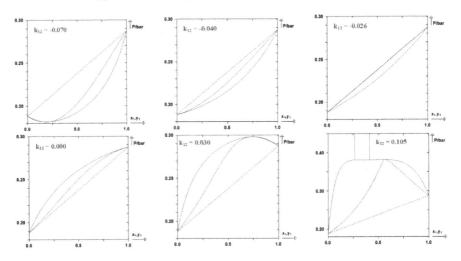

Figure 4.1. *Influence of the parameter k_{12} in the system 2,2,4 trimethylpentane (1) + toluene (2) at 333 K, on the isothermal phase diagram as calculated by the PR equation of state. For a color version of the figure, see www.iste.co.uk/jaubert/ thermodynamic.zip*

How can the binary interaction parameters k_{ij} be estimated? There are three main approaches: either they are **adjusted** in order to reproduce VLE experimental data (this is by far the most reliable method), or they can be **estimated from various correlations**, or they can be **predicted** using group contribution methods.

We will now look at some of the correlations proposed in the literature for the parameter k_{ij}. Most of them are empirical, applicable in very specific situations and cannot be used in extrapolation.

Chueh and Prausnitz[1] proposed a correlation for mixtures of paraffins that, in order to be applied, only requires the critical molar volumes of the pure substances to be known:

$$k_{ij} = 1 - \left(\frac{2\sqrt{v_{c,i}^{1/3} v_{c,j}^{1/3}}}{v_{c,i}^{1/3} + v_{c,j}^{1/3}} \right)^n \qquad [4.15]$$

Stryjek[2] developed a function of the temperature $k_{ij}(T)$ adapted to the SRK equation of state in order to represent mixtures of alkanes. It is of the form:

$$k_{ij} = k_{ij}^0 + k_{ij}^T \left[(T / K) - 273.15 \right] \qquad [4.16]$$

Gao *et al.*[3] proposed a correlation for mixtures of light hydrocarbons (paraffins, naphthenes, aromatics, alkynes) as a function of critical values of pure substances (critical temperature and critical compressibility factor):

$$1 - k_{ij} = \left(\frac{2\sqrt{T_{c,i} T_{c,j}}}{T_{c,i} + T_{c,j}} \right)^{z_{c,ij}} \quad \text{with} \quad z_{c,ij} = \frac{z_{c,i} + z_{c,j}}{2} \qquad [4.17]$$

1 Chueh, P.L. and Prausnitz, J.M. *AIChE J.*,13, 1099–1107 (1967).
2 Stryjek, R. *Fluid Phase Equilib.*, 56, 141–152 (1990).
3 Gao, G. *et al. Fluid Phase Equilib.*, 74, 85–93 (1992).

Kordas *et al.*[4] developed two correlations for binary mixtures of methane + alkane. The first one applies to alkanes that are lighter than eicosane (alkane with 20 carbon atoms), and the second applies to heavier alkanes. In these correlations, ω denotes the acentric factor of the alkane mixed with methane.

$$\begin{cases} k_{ij} = -0.13409\omega + 2.28543\omega^2 - 7.61455\omega^3 + 10.46565\omega^4 - 5.2351\omega^5 \\ k_{ij} = -0.04633 - 0.04367\ln\omega \end{cases}$$

$$[4.18]$$

In the literature, many authors have also worked on proposing binary interaction parameters k_{ij} for mixtures of alkane + CO_2. Experimental results have shown that these k_{ij} diverge greatly from zero and that they are necessary for correct restitution of the thermodynamic properties of these systems. Thus, for this type of mixture, Graboski and Daubert[5] proposed the following correlation developed for the SRK equation:

$$k_{ij} = A + B\left|\delta_i - \delta_j\right| + C\left|\delta_i - \delta_j\right|^2 \qquad [4.19]$$

where δ_i denotes the solubility parameter of component i. The authors have also extended the use of this correlation to mixtures of alkane + N_2 and alkane + H_2S.

Still on the subject of mixtures of CO_2 + alkane, Kato *et al.*[6] developed a temperature function $k_{ij}(T)$ for the PR equation of state:

$$k_{ij} = a(T - b)^2 + c \qquad [4.20]$$

Coefficients a, b, c of this correlation are expressed as functions of the acentric factor of the alkane in question. Moysan *et al.*[7] have applied this same type of correlation to mixtures of {CO_2 or N_2 or H_2 or CO} + alkane.

4 Kordas, A. *et al. Fluid Phase Equilib.* 112, 33–44 (1995).
5 Graboski, M.S. and Daubert, T.E. *Ind. Eng. Chem. Process Des. Dev.*, 17, 448–454 (1978).
6 Kato, K. *et al. Fluid Phase Equilib.*, 7, 219–231 (1981).
7 Moysan, J.M. *et al. Chem. Eng. Science*, 41, 2069–2074 (1986).

Kordas *et al.*[8] also worked on proposing a function $k_{ij}(T)$ for mixtures of CO_2 (denoted as i) + linear alkane (denoted as j):

$$k_{ij} = a(\omega_j) + b(\omega_j) \times T_{r,i} + c(\omega_j) \times T_{r,i}^3 \quad \text{with} : T_{r,i} = \frac{T}{T_{c,i}} \qquad [4.21]$$

Their correlation can also be applied to mixtures containing branched alkanes, aromatic molecules or naphthenic molecules by introducing a modified acentric factor ω_j (the authors provide a correlation of this parameter as a function of the molecular weight and the liquid density of the molecule at 15°C).

Valderrama *et al.*[9,10] worked for their part on the mixtures {CO_2 or N_2 or H_2S) (denoted as i) + light alkane (denoted as j):

$$k_{ij} = A(\omega_j) + B(\omega_j)/T_{r,j} \qquad [4.22]$$

Avlonitis *et al.*[11] worked on the same mixtures as Valderrama *et al.* and proposed the following expression for the PR equation:

$$k_{ij} = Q(\omega_j) - \frac{T_{r,i}^2 + A(\omega_j)}{T_{r,i}^3 + C(\omega_j)} \qquad [4.23]$$

Finally, Nishiumi *et al.*[12] proposed a very general correlation, specific to the PR equation, which can be applied to all mixtures containing paraffins, naphthenes, aromatic molecules, alkynes, CO_2, N_2 and H_2S. In order to be applied, it requires the critical molar volumes of the components to be known:

$$1 - k_{ij} = C + D\frac{v_{c,i}}{v_{c,j}} + E\left(\frac{v_{c,i}}{v_{c,j}}\right)^2 \quad \text{with} : \begin{cases} C = c_1 + c_2|\omega_i - \omega_j| \\ D = d_1 + d_2|\omega_i - \omega_j| \end{cases} \qquad [4.24]$$

8 Kordas, A. *et al. Fluid Phase Equilib.*, 93, 141–166 (1994).
9 Valderrama, J.O. *et al. Fluid Phase Equilib.*, 59, 195–205 (1990).
10 Valderrama, J.O. *et al. Can. J. Chem. Eng.*, 77, 1239–1243 (1999).
11 Avlonitis, G. *et al. Fluid Phase Equilib.*, 101, 53–68 (1994).
12 Nishiumi, H. *et al. Fluid Phase Equilib.*, 42, 43–62 (1988).

Since 2004, we have worked on **predicting** binary interaction parameters $k_{ij}(T)$ for the PR and SRK equations of state using the group-contribution concept. We have named the model formed from the combination of the PR equation of state and our group-contribution method for estimation of $k_{ij}(T)$: PPR78 (for *Predictive Peng–Robinson 1978*). We will now briefly present this model.

The PPR78 model (recently renamed *E*-PPR78, *E* meaning *Enhanced*) uses a predictive method to estimate the $k_{ij}(T)$ in the PR equation. This model is based on prior theoretical work on the prediction of binary parameters of the Van Laar activity coefficient model, but we will not give details of this aspect in this chapter. In order to use the PPR78 model, it is necessary to know for each molecule i the equation of state parameters (a_i, b_i) and structural information like the **group occurrences**. Today the model *E*-PPR78 defines 40 groups (CH_3, CH_2, CH, C, methane, ethane, groups to describe aromatic molecules, to describe naphthenic molecules, to describe permanent gases, to describe CFC gases, to describe mercaptans, water, H_2S, etc.). Group occurrences α_{ik} denote the number of times the group k is present in the molecule i. The coefficient $k_{ij}(T)$ is given by the equation:

$$\begin{cases} k_{ij}(T) = \dfrac{E_{ij}(T) - \left(\dfrac{\sqrt{a_i(T)}}{b_i} - \dfrac{\sqrt{a_j(T)}}{b_j} \right)^2}{2\dfrac{\sqrt{a_i(T) \cdot a_j(T)}}{b_i \cdot b_j}} \\ E_{ij}(T) = -\dfrac{1}{2}\sum_{k=1}^{N_g}\sum_{l=1}^{N_g}(\alpha_{ik} - \alpha_{jk})(\alpha_{il} - \alpha_{jl})A_{kl} \cdot \left(\dfrac{298.15}{T} \right)^{\left(\frac{B_{kl}}{A_{kl}} - 1 \right)} \end{cases} \quad [4.25]$$

In this equation, N_g denotes the total number of groups defined by the method (40 to date), whereas the parameters A_{kl} and B_{kl} are constants that express the interactions between the groups k and l. The complete matrices **A** and **B** are available in the most recent articles (see, for example, the "supporting information" document that features in our article regarding the extension of the model to alkynes[13]).

---

13 Xu, X. *et al. Ind. Eng. Chem. Res.*, 56, 8143–8157 (2017).

To generate a predictive equation for the $k_{ij}(T)$ in the SRK equation, we have established an equation that allows transitions to be made between the $k_{ij}(T)$ in the PR equation and those in the SRK equation. The corresponding model has been named PR$_2$SRK[14].

4.4.3. Advanced mixing rules by combining an equation of state with an activity-coefficient model

We will first present the fundamental idea behind this approach:

An equation of state is capable of providing an expression for the excess Gibbs energy of a multicomponent system (which is denoted as $g^{E,EoS}$ in the following, with *EoS* being the *equation of state*). The analytical expression of $g^{E,EoS}$ that is obtained will thus contain the parameters a_m and b_m of the mixture (see equation [4.10]).

By equating the analytical expression for g^E stemming from the equation of state ($g^{E,EoS}$) to that from an activity-coefficient model (denoted as $g^{E,\gamma}$, this model can be freely chosen among NRTL, UNIQUAC, Wilson, etc.); we can determine the expression for a mixing rule for the parameter a_m / b_m.

There is, however, a problem: the mathematical expression of the activity-coefficient models ($g^{E,\gamma}$) only depends on the temperature and the composition described by the molar fractions of the components in the liquid phase $\mathbf{x} = (x_1,...,x_p)$. It does not depend on the pressure: $g^{E,\gamma}(T,\mathbf{x})$. On the other hand, $g^{E,EoS}$ depends on the temperature, the composition and the total molar volume of the phase: $g^{E,EoS}(T,v,\mathbf{x})$. To equate the two expressions, it is therefore necessary to fix the molar volume in $g^{E,EoS}(T,v,\mathbf{x})$. Similarly, since the equation of state provides a relationship between the pressure and the molar volume [$P(T,v,\mathbf{x})$], we can also fix the pressure.

Let us now look at this approach in more detail. Definition of a mixing rule takes place in **two successive stages**.

---

14 Jaubert, J.N. and Privat, R. *Fluid Phase Equilib.* 224, 285–304 (2010).

Stage 1: the first stage consists of establishing the analytical expression for the molar excess Gibbs energy using a cubic equation of state. We will start with the definition of this quantity:

$$g^E(T,P,\mathbf{x}) = g(T,P,\mathbf{x}) - g^{id}(T,P,\mathbf{x})$$

with : $\begin{cases} g = \text{total molar Gibbs energy of the real phase} \\ g^{id} = \text{total molar Gibbs energy of the phase,} \\ \qquad \text{assumed to be ideal} \end{cases}$ [4.26]

We now define $v^*_{i,liq}(T,P)$, the liquid molar volume of the pure component i at specified temperature and pressure. This molar volume can be obtained by resolving the equation of state for the pure substance i at fixed T and P (see equation [2.47]). We also define $v_{liq}(T,P,\mathbf{x})$, the molar volume of the liquid mixture at specified temperature, pressure and composition. This molar volume can be obtained by solving the equation of state for mixtures at fixed T, P, $\mathbf{x}$ (equation [2.47] is used again here, in which the parameters a and b are those of the mixture, i.e. a_m and b_m).

Once all calculations have been completed, application of equation [4.26] to the cubic equations of state leads to:

$$g^{E,EoS}(T,P,\mathbf{x}) = \overbrace{RT\sum_{i=1}^{p} x_i \ln\left[\frac{v^*_{i,liq}(T,P) - b_i}{v_{liq}(T,P,\mathbf{x}) - b_m(\mathbf{x})}\right]}^{\text{Combinatorial term } a^{E,EoS}_{comb}}$$

$$\left.\begin{array}{l} + \displaystyle\sum_{i=1}^{p} z_i \frac{a_i}{b_i(r_1 - r_2)} \ln\left[\frac{v^*_{i,liq}(T,P) - b_i \cdot r_2}{v^*_{i,liq}(T,P) - b_i \cdot r_1}\right] \\[3ex] - \dfrac{a_m(T,\mathbf{x})}{b_m(\mathbf{x})\cdot(r_1 - r_2)} \ln\left[\dfrac{v_{liq}(T,P,\mathbf{x}) - b_m(\mathbf{x})\cdot r_2}{v_{liq}(T,P,\mathbf{x}) - b_m(\mathbf{x})\cdot r_1}\right] \end{array}\right\} \begin{array}{l}\text{Residual}\\ \text{term}\\ a^{E,EoS}_{res}\end{array}$$ [4.27]

$$+ \underbrace{P\cdot\left[v_{liq}(T,P,\mathbf{x}) - \sum_{i=1}^{p} z_i v^*_{i,liq}(T,P)\right]}_{v^E(T,P,\mathbf{x})}$$

Stage 2: this stage consists of equating the molar excess Gibbs energy given by the equation of state **at a specified reference pressure** (P_{ref}) and the same quantity stemming from the activity-coefficient model:

$$g^{E,EoS}(T,P_{ref},\mathbf{x}) = g^{E,\gamma}(T,\mathbf{x}) \tag{4.28}$$

From equation [4.28], an expression for the parameter a_m / b_m can be deduced.

Three types of advanced mixing rules – selected from among the best known and the most frequently used – are presented below:

– Huron–Vidal mixing rule;

– Wong–Sandler mixing rule;

– MHV ("*modified Huron–Vidal*") mixing rule.

– Huron–Vidal mixing rule: P_{ref}= +∞.

The Huron–Vidal mixing rule considers an infinite reference pressure (the fluid mixture is then in a state of maximum packing and is necessarily liquid). Under infinite pressure, the molar volumes returned by the equation of state become equal to the covolumes:

$$\begin{cases} \lim\limits_{P \to +\infty} v_{liq} = b_m \\ \lim\limits_{P \to +\infty} v_{i,liq}^{*} = b_i \end{cases}$$. The terms of equation [4.27] are then written:

$$\begin{cases} a_{comb}^{E,EoS} = \left[RT\sum\limits_{i=1}^{p} x_i \ln\left(\dfrac{v_{i,liq}^{*} - b_i}{v_{liq} - b_m} \right) \right] \xrightarrow[P \to +\infty]{} 0 \\[2em] a_{res}^{E,EoS} = \left[\sum\limits_{i=1}^{p} x_i \dfrac{a_i}{b_i(r_1 - r_2)} \ln\left(\dfrac{v_{i,liq}^{*} - b_i \cdot r_2}{v_{i,liq}^{*} - b_i \cdot r_1} \right) - \dfrac{a_m}{b_m(r_1 - r_2)} \ln\left(\dfrac{v_{liq} - b_m \cdot r_2}{v_{liq} - b_m \cdot r_1} \right) \right] \\[1em] \qquad \xrightarrow[P \to +\infty]{} C_{EoS}\left[\dfrac{a_m}{b_m} - \sum\limits_{i=1}^{p} x_i \dfrac{a_i}{b_i} \right] \text{ with } \begin{cases} C_{EoS} = \dfrac{1}{r_2 - r_1} \ln\left(\dfrac{1 - r_2}{1 - r_1} \right) \ (\textit{if } r_1 \neq r_2) \\ = \dfrac{1}{r_1 - 1} \ (\textit{if } r_1 = r_2) \end{cases} \\[2em] \dfrac{Pv^E}{RT} = \dfrac{P}{RT}\left(v_{liq} - \sum\limits_i x_i v_{i,liq}^{*} \right) \\[1em] \qquad = \dfrac{P}{RT}\left[(v_{liq} - b_m) - \sum\limits_i x_i(v_{i,liq}^{*} - b_i) + b_m - \sum\limits_i x_i b_i \right] \\[1em] \qquad = \underbrace{\dfrac{P(v_{liq} - b_m)}{RT}}_{\xrightarrow[P \to +\infty]{} 1} - \sum\limits_i x_i \underbrace{\dfrac{P(v_{i,liq}^{*} - b_i)}{RT}}_{\xrightarrow[P \to +\infty]{} 1} + \underbrace{\dfrac{P}{RT}\left(b_m - \sum\limits_i x_i b_i \right)}_{\xrightarrow[P \to +\infty]{} ????} \end{cases} \tag{4.29}$$

In other words, considering an infinite reference pressure, equation [4.27] can be rewritten as:

$$g^{E,EoS}(T, P = +\infty, \mathbf{x}) = 0 + C_{EoS}\left[\frac{a_m}{b_m} - \sum_{i=1}^{p} x_i \frac{a_i}{b_i}\right] + 1 - 1 + \underbrace{\frac{P}{RT}\left(b_m - \sum_i x_i b_i\right)}_{P \to +\infty} \qquad [4.30]$$

The last term in equation [4.30] has an undetermined limit at infinite pressure because the value of the term $\left(b_m - \sum_i x_i b_i\right)$ is unknown. If this value is finite and not zero or infinite, the last term of equation [4.30] diverges. The only solution to avoid this divergence is to set it equal to zero. The Huron–Vidal mixing rule imposes a **linear mixing rule of the composition for the covolume:**

$$\text{Huron and Vidal impose :} \qquad b_m = \sum_i x_i b_i \qquad [4.31]$$

From equations [4.28], [4.30] and [4.31], we arrive at:

$$\begin{cases} g^{E,\gamma}(T,\mathbf{x}) = C_{EoS}\left[\dfrac{a_m}{b_m} - \sum_{i=1}^{p} x_i \dfrac{a_i}{b_i}\right] \\ b_m = \sum_i x_i b_i \end{cases} \qquad [4.32]$$

where the expression for C_{EoS} is given in equation [4.29]. By rearranging equation [4.32], we obtain:

Huron-Vidal mixing rules:

$$\begin{cases} \dfrac{a_m}{b_m} = \sum_{i=1}^{p} x_i \dfrac{a_i}{b_i} + \dfrac{g^{E,\gamma}(T,\mathbf{x})}{C_{EoS}} \quad \text{with} \quad \begin{cases} C_{EoS} = \dfrac{1}{r_2 - r_1}\ln\left(\dfrac{1-r_2}{1-r_1}\right) \ (if \ r_1 \neq r_2) \\[2mm] = \dfrac{1}{r_1 - 1} \ (if \ r_1 = r_2) \end{cases} \\ b_m = \sum_i x_i b_i \end{cases}$$

$$[4.33]$$

We note that under infinite pressure, the combinatorial term from the equation of state ($a_{comb}^{E,EoS}$) cancels out. Only the residual (enthalpic) part remains. Excellent results will be obtained if we select for $g^{E,\gamma}$: the NRTL model that only contains a residual part (see Chapter 3), the residual part of the UNIQUAC or Wilson models.

– Wong–Sandler mixing rules (inspired by Huron-Vidal rules).

In the hope of improving the performances of the Huron–Vidal mixing rules, Wong and Sandler proposed to re-examine them. Reasoning not on the basis of the excess Gibbs energy but on the excess Helmholtz energy, they demonstrated that only the constraint on the ratio a_m/b_m should be conserved and the mixing rule for the parameter b_m is not necessarily linear.

To find the mixing rule for b_m, they started from the observation that a linear mixing rule does not allow the second virial coefficient B to be expressed as a quadratic function of the composition, despite the fact that this form is imposed by statistical thermodynamics:

$$\left\{ \begin{array}{l} \text{Form of B predicted by statistical thermodynamics} = \sum_{i=1}^{p}\sum_{j=1}^{p} x_i x_j B_{ij} \\[2em] \text{Expression for B predicted by cubic EoS: } B = b_m - \dfrac{a_m}{RT} \end{array} \right. \qquad [4.34]$$

To satisfy this constraint associated with the second virial coefficient, they therefore propose the following mixing rule:

$$\left\{ \begin{array}{l} \text{Wong-Sandler mixing rules:} \\[1em] \dfrac{a_m}{b_m} = \sum_{i=1}^{p} x_i \dfrac{a_i}{b_i} + \dfrac{g^{E,\gamma}(T,\mathbf{x})}{C_{EoS}} \\[2em] b_m(T,\mathbf{x}) = \dfrac{\displaystyle\sum_{i=1}^{p}\sum_{j=1}^{p} x_i x_j \dfrac{\left(b_i - \dfrac{a_i}{RT}\right) + \left(b_j - \dfrac{a_j}{RT}\right)}{2}(1-k_{ij})}{1 - \dfrac{1}{RT}\left[\displaystyle\sum_{i=1}^{p} x_i \dfrac{a_i(T)}{b_i} + \dfrac{g^{E,\gamma}(T,\mathbf{x})}{C_{EoS}}\right]} \\[3em] C_{EoS} = \dfrac{1}{r_2 - r_1} \cdot \ln\left(\dfrac{1-r_2}{1-r_1}\right) \text{[SRK or PR EoS]} \end{array} \right. \qquad [4.35]$$

The binary interaction parameter k_{ij} is an adjustable parameter.

– **MHV mixing rule:** $P_{ref} = 0$.

This mixing rule was proposed by Michelsen, a Danish researcher, at the beginning of the 1990s. By introducing the following quantities:

$$\left\{ \begin{array}{l} \eta_{liq} = \dfrac{b_m}{v_{liq}} : \text{packing fraction of the mixture at } (T, P, \mathbf{x}) \\[12pt] \eta_{liq,i} = \dfrac{b_i}{v_{liq,i}} : \text{packing fraction of the pure substance } i \text{ at } (T, P) \\[12pt] \Gamma = \dfrac{a_m}{b_m RT} \text{ and } \Gamma_i = \dfrac{a_i}{b_i RT} \\[12pt] Q(\Gamma, \eta_{liq}) = -\ln\left(\dfrac{1 - \eta_{liq}}{\eta_{liq}}\right) - \dfrac{\Gamma}{r_1 - r_2} \ln\left(\dfrac{1 - \eta_{liq} r_2}{1 - \eta_{liq} r_1}\right) \end{array} \right. \qquad [4.36]$$

equation [4.27] is rewritten as:

$$\frac{g^{E,EoS}}{RT} = Q(\Gamma, \eta_{liq}) - \sum_i x_i Q(\Gamma_i, \eta_{liq,i}) + \sum_i x_i \ln\left(\frac{b_i}{b_m}\right) + \frac{Pv^E}{RT} \qquad [4.37]$$

With the objective of equating the excess Gibbs energies of the activity-coefficient model and of the equation of state at $P_{ref} = 0$, we seek to express $g^{E,EoS}$ under zero pressure. Beforehand, it should be mentioned that at zero pressure, the liquid molar volume (metastable) predicted by the equation of state only depends on Γ and on the universal constants of the equation of state. These characteristics also apply to the liquid packing fraction at zero pressure predicted by the equation of state. Effectively, at zero pressure, the cubic equation of state for mixtures is written as:

$$P = 0 = \frac{RT\eta}{b_m(1 - \eta)} - \frac{a_m \eta^2}{b_m^2(1 - \eta r_1)(1 - \eta r_2)}$$

$$= \frac{RT\eta}{b_m(1 - \eta)} - \frac{RT\Gamma\eta^2}{b_m(1 - \eta r_1)(1 - \eta r_2)} \qquad [4.38]$$

From this expression, we can deduce (by solving a polynomial equation) the expression for the liquid packing fraction at zero pressure:

$$\eta_0(\Gamma) = \frac{r_1 + r_2 + \Gamma + \sqrt{(r_1 + r_2 + \Gamma)^2 - 4(r_1 r_2 + \Gamma)}}{2(r_1 r_2 + \Gamma)} \qquad [4.39]$$

Consequently, at zero pressure, the function Q defined by equation [4.36] depends only on Γ. We write $Q_0(\Gamma) = Q(\Gamma, \eta_0(\Gamma))$ in the following. We will now express $g^{E,EoS}$ at zero pressure.

$$\frac{g^{E,EoS,P=0}}{RT} = Q_0(\Gamma) - \sum_i x_i Q_0(\Gamma_i) + \sum_i x_i \ln\left(\frac{b_i}{b_m}\right)$$

$$\text{with } Q_0(\Gamma) = -\ln\left(\frac{1 - \eta_0(\Gamma)}{\eta_0(\Gamma)}\right) - \frac{\Gamma}{r_1 - r_2} \ln\left(\frac{1 - \eta_0(\Gamma) \cdot r_2}{1 - \eta_0(\Gamma) \cdot r_1}\right) \qquad [4.40]$$

Equating the right-hand side of equation [4.40] to the expression given by an activity-coefficient model, we obtain:

MHV mixing rule:

$$\frac{g^{E,\gamma}(T,\mathbf{x})}{RT} = Q_0(\Gamma) - \sum_i x_i Q_0(\Gamma_i) + \sum_i x_i \ln\left(\frac{b_i}{b_m}\right) \qquad [4.41]$$

Equation [4.41] defines the MHV mixing rule (for "modified Huron–Vidal"). Solving this equation allows the value of the parameter $\Gamma = a_m/(b_m RT)$ to be deduced. Effectively, if the temperature T and the composition $\mathbf{x}$ of the mixture are fixed, it is possible to estimate $\Gamma_i = a_i(T)/(b_i RT)$. If in addition, a mixing rule for parameter $b_m(\mathbf{x})$ is selected (we will note that the mixing rule on this parameter is free, MHV does not impose it), equation [4.41] reads as an equation with one unknown Γ that can be resolved numerically.

This method has two main disadvantages:

– on the one hand, the procedure is computationally heavy: equation [4.41] must be solved for each value of the couple $(T, \mathbf{x})$, which is costly in terms of calculation time;

– on the other hand, equation [4.41] is not always soluble. For certain values of Γ or Γ_i, the function Q_0 is not defined.

In order to overcome these problems, the MHV-1 and MHV-2 mixing rules were proposed.

– *MHV-1 mixing rule (derivative of MHV)*

Noting that the representative curve of the function Q_0 is a straight line (as shown in Figure 4.2), Michelsen proposed the following approximations:

$$\begin{cases} Q_0(\Gamma) \approx q_0 + q_1 \cdot \Gamma \\ Q_0(\Gamma_i) \approx q_0 + q_1 \cdot \Gamma_i \end{cases} \qquad [4.42]$$

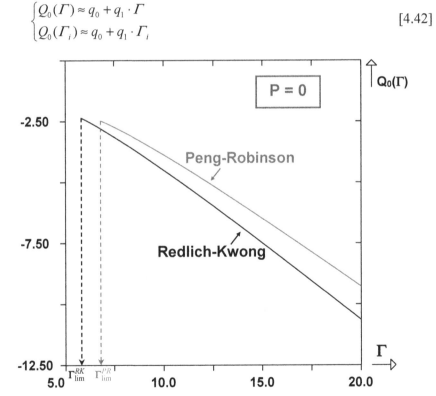

Figure 4.2. *Form of the function Q_0 defined by equation [4.40]. For a color version of the figure, see www.iste.co.uk/jaubert/thermodynamic.zip*

The parameters q_0 and q_1 depend on the equation of state in question. They can be determined by simple linear regression on curves that represent Q_0. For Van der Waals (VdW), RK and PR equations, they are:

$$
\begin{cases}
\text{PR}: & q_0 = 1.527 \text{ and } q_1 = -0.534 \\
\text{SRK}: & q_0 = 1.581 \text{ and } q_1 = -0.604 \\
\text{VdW}: & q_0 = 2.197 \text{ and } q_1 = -0.900
\end{cases}
\qquad [4.43]
$$

Combining equations [4.41] and [4.42], we arrive at the mixing rule:

$$
\frac{g^{E,\gamma}(T,\mathbf{x})}{RT} = q_1 \left(\Gamma - \sum_i x_i \Gamma_i \right) + \sum_i x_i \ln\left(\frac{b_i}{b_m} \right) \qquad [4.44]
$$

The advantage of this formulation is that it gives an explicit expression for Γ, thus removing the resolution problems presented by the MHV model. By rearranging equation [4.44], we obtain:

$$
\boxed{
\begin{aligned}
&\text{MHV-1 mixing rule:} \\
&\begin{cases}
\Gamma = \dfrac{a_m}{b_m RT} = \displaystyle\sum_{i=1}^{p} x_i \dfrac{a_i}{b_i RT} + \dfrac{\dfrac{g^{E,\gamma}}{RT} - \displaystyle\sum_i x_i \ln\left(\dfrac{b_i}{b_m} \right)}{q_1} \\
b_m : \text{free choice}
\end{cases}
\end{aligned}
} \qquad [4.45]
$$

As an example, the PSRK (*Predictive*-SRK) model is based on the use of these mixing rules (with a value of q_1 modified by the authors). It is written as:

$$
\text{PSRK:}
\begin{cases}
\dfrac{a_m}{b_m} = \displaystyle\sum_{i=1}^{p} x_i \dfrac{a_i}{b_i} + \dfrac{g^{E,\gamma}_{\text{UNIFAC}} - RT \displaystyle\sum_i x_i \ln\left(\dfrac{b_i}{b_m} \right)}{q_1} \\
b_m = \displaystyle\sum_i x_i b_i \\
q_1 = -0.64663
\end{cases}
\qquad [4.46]
$$

– MHV-2 mixing rule (derivative of MHV)

Following the same principle as MHV-1, the function Q_0 can be approximated by a second-degree polynomial function:

$$\begin{cases} Q_0(\Gamma) \approx q_0 + q_1 \cdot \Gamma + q_2 \cdot \Gamma^2 \\ Q_0(\Gamma_i) \approx q_0 + q_1 \cdot \Gamma_i + q_2 \cdot \Gamma_i^2 \end{cases}$$

[4.47]

Thus, combining equations [4.41] and [4.47], we obtain:

$$\boxed{\begin{aligned} &\text{MHV-2 mixing rule:} \\ &\begin{cases} q_1\left(\Gamma - \sum_i x_i \Gamma_i\right) + q_2\left(\Gamma^2 - \sum_i x_i \Gamma_i^2\right) + \sum_i x_i \ln\left(\dfrac{b_i}{b_m}\right) - \dfrac{g^{E,\gamma}(T,\mathbf{x})}{RT} = 0 \\ b_m : \text{free choice} \end{cases} \end{aligned}}$$

[4.48]

This said, Γ is a solution of the second-degree polynomial equation [4.48].

NOTE.– MHV mixing rules have been widely used since their initial proposal. However, over the last 15 years or so, various studies have shown that the presence of two types of combinatorial terms in the mixing rule, i.e. from the $g^{E,\gamma}(T,\mathbf{x})$ model and from the equation of state, which are of the same type but are not quantitatively equivalent, penalizes the model. This phenomenon is referred to as the ***presence of a double combinatorial term***. The authors of the UMR-PRU and VTPR models thus recommend that only the ***residual parts*** of g^E from an activity-coefficient model and from the equation of state should be identified. The MHV-1 mixing rule is then given by equation [4.49]

$$\boxed{\begin{aligned} &\text{MHV-1 mixing rule without the problem of the double} \\ &\qquad\qquad\text{combinatorial term:} \\ &\begin{cases} \Gamma = \dfrac{a_m}{b_m RT} = \displaystyle\sum_{i=1}^{p} x_i \dfrac{a_i}{b_i RT} + \dfrac{\left(\dfrac{g_{r\acute{e}s}^{E,\gamma}}{RT}\right)}{q_1} \\ b_m : \text{free choice} \end{cases} \end{aligned}}$$

[4.49]

4.4.4. Short summary of the mixing rules for cubic equations of state

– Classic mixing rules

Strengths:

– they are easy to implement;

– they lead to excellent results when all components have low polarity and are non-associating;

– the use of a binary interaction parameter k_{ij}, which depends on the temperature T, greatly improves the results;

– many different methods can be used for estimating k_{ij}.

Weaknesses:

– they are not suitable for modeling components with average to strong polarity;

– they are not suitable for modeling mixtures involving associating molecules (water, alcohols).

– Advanced mixing rules (defined from an activity-coefficient model).

Strengths:

– they allow complex systems containing polar molecules to be represented;

– the use of the residual part of the Wilson, UNIQUAC and NRTL models in the mixing rules produces results that are accurate and of equivalent quality.

Weaknesses:

– parameters of the g^E model must be fitted to experimental data;

– they are more difficult to program than classic mixing rules;

– significant differences can persist between experimental behaviors and those predicted by the model for systems that contain associating molecules (water, alcohols).

4.5. Equations of state based on the "statistical associating fluid theory" (SAFT)

In the same way as the SAFT equations of state for pure substances, the equations of state for mixtures are written in the general form:

$$\begin{cases} a = a^{\bullet} + a^{res,HS} + a^{res,Disp\,seg} + a^{res,Chain\,of\,seg} \\ \quad + a^{res,Assoc} + a^{res,Polar} + ... \\ P(T,v,z) = -\left(\dfrac{\partial a}{\partial T}\right)_{v,z} \end{cases}$$ [4.50]

SAFT equations are extended to mixtures by using the results of statistical thermodynamics applied to hard spheres and the one-fluid theory for the dispersive term. As a reminder, the terms of the equation of state for pure substances that correspond to the repulsive and dispersive effects require three input parameters to be known: m (number of segments), σ (diameter of a segment) and ε (depth of sink of potential). The number of segments of a mixture is given by a linear mixing rule:

$$m(\mathbf{z}) = \sum_i z_i \cdot m_i$$ [4.51]

In the same way as for cubic equations of state, it is necessary to define combining rules for the crossover parameters of the parameters σ and ε. If i and j denote two segments, we generally use:

Lorentz-Berthelot combining rules:

$$\begin{cases} \varepsilon_{ij} = \sqrt{\varepsilon_{ii}\varepsilon_{jj}}\,(1-k_{ij}) \\ \sigma_{ij} = \dfrac{\sigma_{ii}+\sigma_{jj}}{2}(1-l_{ij}) \end{cases}$$ [4.52]

where k_{ij} and l_{ij} are binary interaction parameters that are adjustable or can be predicted by correlations. Currently, most process simulators do not include an association term in their SAFT models. For this reason, we limit the discussion to the non-associative terms of these models.

To achieve a good representation of the mixtures, it is not rare – in practice (and although certain SAFT model developers advise against it)

– for a binary interaction parameter to be added in the model (often k_{ij}).

For example, Figure 4.3 shows the phase diagrams of the system $CO_2(1)$ + ethane(2) that have been modeled using the PC-SAFT equation (without an association term, nor a polarity term) in the case where k_{12} is zero and in the case where the same parameter is equal to 0.10. Another example, with the system composed of $CO_2(1)$ + p-cymene(2), is given in Figure 4.4.

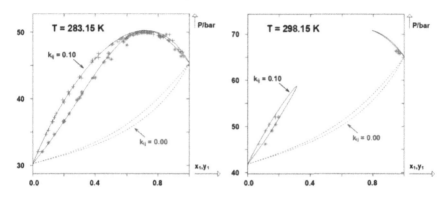

Figure 4.3. *Modeling of the system $CO_2(1)$ + ethane(2) by the PC-SAFT equation (without association or polarity terms but with the use of a binary interaction parameter). For a color version of the figure, see www.iste.co.uk/jaubert/ thermodynamic.zip*

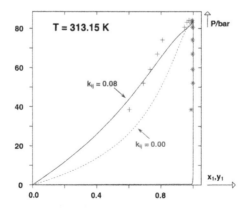

Figure 4.4. *Modeling of the system $CO_2(1)$ + p-cymene(2) using the PC-SAFT equation (without association or polarity terms but with the use of a binary interaction parameter). For a color version of the figure, see www.iste.co.uk/jaubert/ thermodynamic.zip*

We will mention here the existence of correlations proposed in the literature to estimate k_{ij} parameters. These correlations generally express the parameters as a function of a property that can be experimentally determined: the ionization energy (denoted as I_i for a component i). Equation [4.53] gives some examples.

$$\left\{ \begin{array}{l} \text{Hudson - Mac Coubrey: } k_{ij} = 1 - \left[2^7 \left(\frac{\sqrt{I_i I_j}}{I_i + I_j} \right) \left(\frac{\sigma_i^3 \sigma_j^3}{(\sigma_i + \sigma_j)^6} \right) \right] \\ \\ \text{GC-PPC-SAFT model: } k_{ij} = 1 - \left(\frac{2\sqrt{I_i I_j}}{I_i + I_j} \right) \end{array} \right. \qquad [4.53]$$

We will now compare the SAFT equations of state and the cubic equations of state in terms of advantages and disadvantages:

Cubic equations of state:

– they produce better restitutions of critical points;

– they have weaknesses for systems of strongly associating molecules;

– they can be combined with many methods that allow the binary interaction parameters (k_{ij} for the classic mixing rules or $g^{E,\gamma}$ parameters for advanced mixing rules) to be estimated. As an example, PPR78 equation can be used for classic mixing rules and UNIFAC can be used for advanced mixing rules.

– they are simple to program and parameterize.

SAFT equations:

– they produce better restitutions of the densities of mixtures (unless the technique of volumetric translation is incorporated into the cubic equation of state);

– they are highly suitable for systems that contain polymers or heavy molecules;

– they provide a good correlation for systems that contain associating molecules (on the condition that the many parameters of the association term are known);

− they are difficult to parameterize as the parameters are not easily available.

4.6. Specific equations of state for particular mixtures

This type of equation was presented in section 2.3.5 in the context of pure substances. They also exist for certain specific mixtures.

The NIST (National Institute of Standards and Technology) develops a software known as REFPROP (Reference Fluid Thermodynamic and Transport Properties Database), which includes specific equations of state for mixtures of fixed global composition. For example, the model GERG-2008 is applicable to natural gases, to the air, to mixtures of CFC gases and so on.

These specific equations of state contain many parameters, but they are − on the other hand − extremely accurate.

4.7. Practical use of the equations of state for mixtures

4.7.1. *What can be calculated from an equation of state for mixtures?*

Calculation of state properties

In the same way as an equation of state for pure substances, the equation of state for mixtures allows all the properties of a mixture to be calculated (enthalpy, entropy, exergy, heat capacities, etc.). More precisely, an equation of state only gives an estimate of the correction that must be applied to a mixture of perfect gases in order to obtain that of the real mixture. We will therefore write:

$$\underbrace{x}_{\substack{\text{property of}\\\text{real mixture}}} = \underbrace{x^{\bullet}}_{\substack{\text{property of the}\\\text{perfect gas mixture}}} + \underbrace{x^{correction}}_{\substack{\text{given by the}\\\text{equation of state}\\\text{for the real mixture}}} \qquad [4.54]$$

Since the equation of state can also be applied to pure substances, it allows all the mixing properties to be calculated (e.g. the enthalpy of mixing h^M, the heat capacity of mixing c_p^M, etc.).

Vapor–liquid equilibrium

When the equation of state is applicable to the liquid and vapor phases, the conditions of equilibrium between phases can be solved and the phase diagrams can then be calculated.

4.7.2. Calculation of mixing properties using a pressure-explicit equation of state

DEFINITION.–By definition, a **mixing property** (also called property change on mixing) is the change of this property that accompanies creation of the mixture from individual pure substances taken in their stable state at the same temperature and the same pressure.

This definition is illustrated in Figure 4.5.

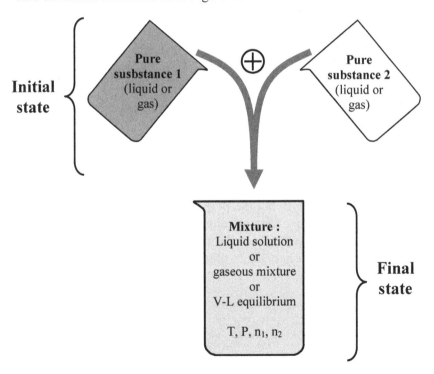

Figure 4.5. *Illustration of the definition of a mixing property*

For example, the (molar) enthalpy, entropy and heat capacity of mixing are given by:

$$
\begin{cases}
h^M(T,P,\mathbf{x}) = \underbrace{h(T,P,\mathbf{x})}_{\substack{\text{total molar enthalpy}\\\text{of the mixture}}} - \sum_i x_i \underbrace{h^*_{i,\text{stable state}}(T,P)}_{\substack{\text{molar enthalpy of}\\\text{pure susbstance } i}} \\[2em]
s^M(T,P,\mathbf{x}) = \underbrace{s(T,P,\mathbf{x})}_{\substack{\text{total molar entropy}\\\text{of the mixture}}} - \sum_i x_i \underbrace{s^*_{i,\text{stable state}}(T,P)}_{\substack{\text{molar entropy of}\\\text{pure susbstance } i}} \\[2em]
c_P^M(T,P,\mathbf{x}) = \underbrace{c_P(T,P,\mathbf{x})}_{\substack{\text{total molar heat capacity}\\\text{of the mixture}}} - \sum_i x_i \underbrace{c^*_{P,i,\text{stable state}}(T,P)}_{\substack{\text{molar heat capacity of}\\\text{pure susbstance } i}}
\end{cases}
\qquad [4.55]
$$

We will now introduce the residual-TV properties into the expressions of molar enthalpy for the mixture and for the pure substances:

$$
\begin{cases}
h(T,P,\mathbf{x}) = h^*(T,\mathbf{x}) + h^{res-TV}(T,P,\mathbf{x}) = \sum_i x_i h_i^*(T) + h^{res-TV}(T,P,\mathbf{x}) \\[1em]
h_i^*(T,P) = h_i^*(T) + h_i^{res-TV}(T,P)
\end{cases}
\qquad [4.56]
$$

The enthalpy of mixing is then written as:

$$
h^M(T,P,\mathbf{x}) = h^{res-TV}(T,P,\mathbf{x}) - \sum_i x_i h_i^{res-TV}(T,P)
\qquad [4.57]
$$

By applying a similar procedure, we show that the entropy and the calorific capacity of mixing are given by:

$$
\begin{cases}
s^M(T,P,\mathbf{x}) = s^{res-TV}(T,P,\mathbf{x}) - \sum_i x_i s_i^{res-TV}(T,P) - R\sum_i x_i \ln x_i \\[1em]
c_P^M(T,P,\mathbf{x}) = c_P^{res-TV}(T,P,\mathbf{x}) - \sum_i x_i c_{P,i}^{res-TV}(T,P)
\end{cases}
\qquad [4.58]
$$

We will now focus specifically on calculation of the molar enthalpy of mixing, given by equation [4.57]. The residual-TV properties are expressed naturally via the set of variables (temperature, molar volume), and equation [4.57] is rewritten in the form:

$$h^M(T,P,\mathbf{x}) = h^M(T,v,\mathbf{x}) = h^{res-TV}(T,v,\mathbf{x}) - \sum_i x_i h_i^{res-TV}(T,v_i^*) \qquad [4.59]$$

where v denotes the molar volume of the mixture in the $(T,P,\mathbf{x})$ state, whereas v_i^* denotes the molar volume of the pure substance i in the (T,P) state.

Generally, as seen in Chapter 2 in the case of pressure-explicit equations of state, the residual-TV molar entropy, enthalpy and heat capacity at constant pressure of a mixture are given by the same expressions as for pure substances (see equation [2.41]):

$$
\begin{cases}
s^{res-TV} = -\left(\dfrac{\partial a^{res-TV}}{\partial T}\right)_{v,\mathbf{x}} \\[2em]
h^{res-TV} = a^{res-TV} + T \cdot s^{res-TV} + v \cdot P^{res-TV} = a^{res-TV} - T\left(\dfrac{\partial a^{res-TV}}{\partial T}\right)_{v,\mathbf{x}} \\[1em]
\qquad\qquad\qquad\qquad\qquad\qquad\qquad\qquad\qquad + P(T,v,\mathbf{x}) \cdot v - RT \\[2em]
c_P^{res-TV} = \left(\dfrac{\partial u^{res-TV}}{\partial T}\right)_{v,\mathbf{x}} - R - T\left[\left(\dfrac{\partial P(T,v,\mathbf{x})}{\partial T}\right)_{v,\mathbf{x}}\right]^2 \bigg/ \left(\dfrac{\partial P(T,v,\mathbf{x})}{\partial v}\right)_{T,\mathbf{x}} \\[2em]
\text{with}: a^{res-TV} = -\displaystyle\int_{+\infty}^{v}\left[P(T,v,\mathbf{x}) - \dfrac{RT}{v}\right]dv
\end{cases}
$$

$$[4.60]$$

In the case of cubic equations of state, the residual-TV molar enthalpy of the mixture is given by:

$$h^{res-TV}(T,v,\mathbf{x}) = \frac{1}{b_m(\mathbf{x}) \cdot (r_1 - r_2)}\left[a_m(T,\mathbf{x}) - T \cdot \left(\frac{\partial a_m}{\partial T}\right)_{\mathbf{x}}\right]$$

$$\ln\left[\frac{v - b_m(\mathbf{x}) \cdot r_1}{v - b_m(\mathbf{x}) \cdot r_2}\right] + \frac{RTb_m(\mathbf{x})}{v - b_m(\mathbf{x})} - \frac{a_m(T,\mathbf{x}) \cdot v}{\left[v - b_m(\mathbf{x}) \cdot r_1\right]\left[v - b_m(\mathbf{x}) \cdot r_2\right]}$$

$$[4.61]$$

For a pure substance, the expression is deduced from that for a mixture by writing:

$$h_i^{res-TV}(T,v_i^*) = h^{res-TV}(T,v,x_i = 1) \qquad [4.62]$$

NOTE.– In the equations of state for mixtures, correct restitution of enthalpic data or heat capacity data is determined by dependence on temperature of the binary interaction parameters. To illustrate this fact, we will consider an equimolar mixture of n-hexane(1) + n-decane(2) at 20°C and at atmospheric pressure, modeled by the PR equation of state using Van der Waals mixing rules. If we select a constant binary interaction parameter of $k_{12} = -0.0003657$, the molar mixing enthalpy is $h^M = 3.90 \; \text{J} \cdot \text{mol}^{-1}$; on the other hand, if we choose to express the binary interaction coefficient k_{12} using a function of the temperature:

$$\begin{cases} A = -0.00207 \\ B = 0.2214 \\ C = 84.0301 \end{cases} \quad \text{and} \quad k_{12} = A + \frac{B}{T} + \frac{C}{T^2} = -0.0003657, \quad \text{we} \quad \text{obtain:}$$

$h^M = 36.4 \; \text{J} \cdot \text{mol}^{-1}$.

4.7.3. Calculation principle of vapor–liquid equilibrium using the φ-φ approach

In the φ-φ approach, the vapor–liquid equilibrium conditions are written as:

$$\begin{cases} \boxed{x_i \cdot \varphi_{i,liq}(T, v_{liq}, \mathbf{x}) = y_i \cdot \varphi_{i,vap}(T, v_{vap}, \mathbf{y})} \quad \forall i \in \{1; ...; p\} \\ P(T, v_{liq}, \mathbf{x}) = P(T, v_{vap}, \mathbf{y}) \end{cases} \quad [4.63]$$

The fugacity coefficients φ_i are determined using a pressure-explicit equation of state, i.e. an equation of state capable of representing both the liquid AND vapor phases (e.g. a cubic or a SAFT-type equation of state) according to:

$$\ln \varphi_i(T, v, \mathbf{z}) = -\ln \left(\frac{P \cdot v}{RT} \right) - \int_{+\infty}^{V} \left[\left(\frac{\partial P/(RT)}{\partial n_i} \right)_{T, V, n_{j \neq i}} - \frac{1}{V} \right] \cdot dV \quad [4.64]$$

We should recall that in the case of a system with p components in

VLE (the number of coexisting phases is $\varphi = 2$), the Gibbs phase rule stipulates that the variance is:

$$v = p + 2 - \varphi = p + 2 - 2 = p \qquad [4.65]$$

Thus, the Gibbs variance of a two-phase system is p (number of components). It is therefore necessary to specify p intensive phase variables (among $T, P, x_1, ..., x_p, y_1, ..., y_p, v_{liq}, v_{vap}$) in order to carry out a VLE calculation, i.e. to determine the intensive state of the system phases in VLE. We will now focus in more detail on the case of a binary system ($p = 2$).

In this case, the eight variables in the problem are:

– the temperature T;

– the pressure P;

– the total molar volumes of the liquid and vapor phases: v_{liq} and v_{vap};

– the molar fractions of the components in the liquid phase: (x_1, x_2);

– the molar fractions of the components in the vapor phase: (y_1, y_2).

The six equations that link these eight variables together are:

- The equilibrium conditions:

$$\left[\begin{array}{l} x_1 \cdot \varphi_{1,liq}(T, v_{liq}, x_1, x_2) = y_1 \cdot \varphi_{1,vap}(T, v_{vap}, y_1, y_2) \\ x_1 \cdot \varphi_{1,liq}(T, v_{liq}, x_1, x_2) = y_1 \cdot \varphi_{1,vap}(T, v_{vap}, y_1, y_2) \end{array} \right.$$

- In each phase, the EoS relates the variables $P, v, T, composition$:

$$\left[\begin{array}{l} P = P_{EoS}(T, v_{liq}, x_1, x_2) \\ P = P_{EoS}(T, v_{vap}, y_1, y_2) \end{array} \right.$$

- 2 closure equations:

$$\left[\begin{array}{l} x_1 + x_2 = 1 \\ y_1 + y_2 = 1 \end{array} \right.$$

$$[4.66]$$

The variance, which is also interpreted as the difference between the number of variables and the number of equations, is equal to 2. It is therefore necessary to specify two independent variables so that the number of unknowns in the problem is equal to the number of equations (condition that is necessary but not sufficient in order for the VLE calculation to yield a solution).

There are various possible combinations of specified variables/calculated variables as has previously been illustrated in Table 3.2. Table 4.1 illustrates the main combinations when a VLE is calculated using an pressure-explicit equation of state.

We specify	We calculate	Name of calculation
T and x_1	$P, x_2, y_1, y_2, v_{liq}, v_{vap}$	Bubble-point pressure
T and y_1	$P, x_1, x_2, y_2, v_{liq}, v_{vap}$	Dew-point pressure
P and x_1	$T, x_2, y_1, y_2, v_{liq}, v_{vap}$	Bubble-point temperature
P and y_1	$T, x_1, x_2, y_2, v_{liq}, v_{vap}$	Dew-point temperature

Table 4.1. *Four main types of VLE calculations for binary systems modeled with a pressure-explicit equation of state*

To remain concise, we will not go into further details of VLE calculations using equations of state. This would be a worthy subject for an entirely separate chapter.

5

General Summary: Decision Tree to Select a Thermodynamic Model in Order to Simulate or Design a Chemical Process

In this chapter, which serves as a conclusion to this book, we go back over the various aspects of selection of thermodynamic models that have been described in the previous chapters. The information here is presented as a summary, mainly in the form of decision trees derived from the information given in Chapters 1–4.

5.1. Selection of thermodynamic models for representation of pure substances

Preliminary questions to answer before selecting a model:

– What is my objective? (i.e., what are the properties that the model must be able to estimate?)

– What are the domains of interest for temperature and pressure?

– What are the structural and chemical characteristics of the pure substance?

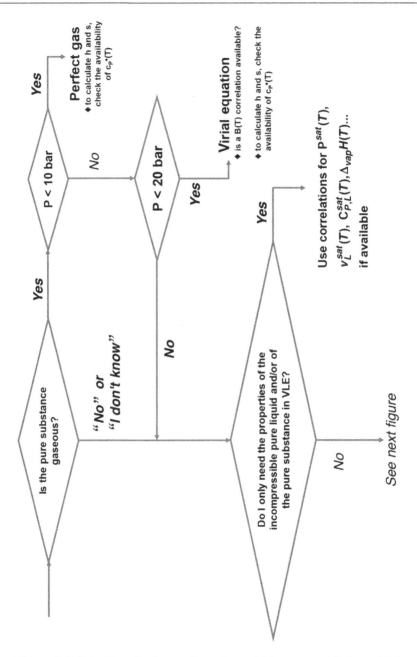

Figure 5.1. *Selection of a thermodynamic model for a pure substance (1/2)*

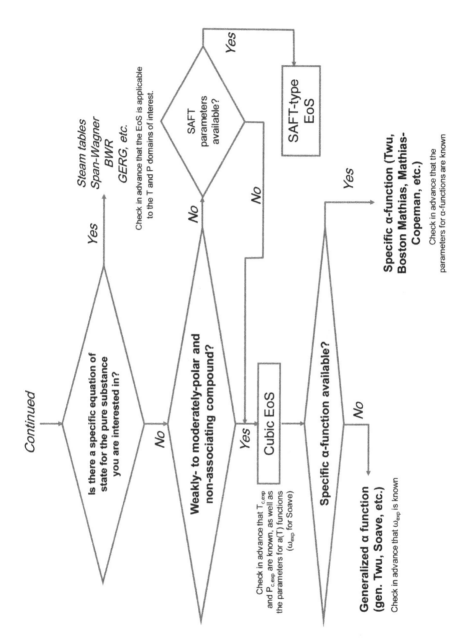

Figure 5.2. *Selection of a thermodynamic model for a pure substance (2/2)*

We recall that it is possible to **translate molar volumes of cubic equations of state** to improve the restitution of liquid densities. The values of the translations depend on:

– the equation of state used (SRK, VdW, PR, etc.);

– the compound in question.

If the liquid densities are a property of interest for you, then the volume translation can simply be estimated. Ideally, if you have an experimental value for the liquid molar volume $v_{liq,\exp}^{sat}(T)$ [or for the liquid density $\rho_{liq,\exp}^{sat}(T) = (\text{Molecular weight}) / v_{liq,\exp}^{sat}(T)$] at a given temperature, you need to calculate it using:

$$c = v_{L,EoS\ untranslated}^{sat}(T) - v_{L,\exp}^{sat}(T) \qquad [5.1]$$

where $v_{L,EoS\ untranslated}^{sat}$ denotes the molar volume predicted by the untranslated equation of state. Otherwise, a correlation must be used (often, the Péneloux correlation that depends on the Rackett z_{RA} compressibility factor; it is therefore necessary to ensure availability of z_{RA} in advance).

Another solution (less elegant but highly effective): bypassing the equation of state to calculate the liquid densities and use available correlations for $v_{L}^{sat}(T)$ (this solution is often applied by default in commercial process simulators).

5.2. Selection of thermodynamic models for representation of mixtures

Preliminary questions to answer before selecting a model:

– What is my objective? (i.e. what are the properties that the model must be capable of estimating?)

– What are the domains of interest for temperature, pressure and composition?

– What are the structural and chemical characteristics of the components of the mixture?

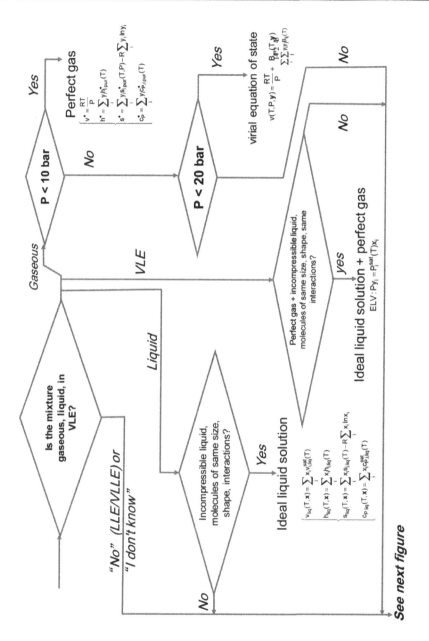

Figure 5.3. *Selection of a thermodynamic model for a mixture (1/4)*

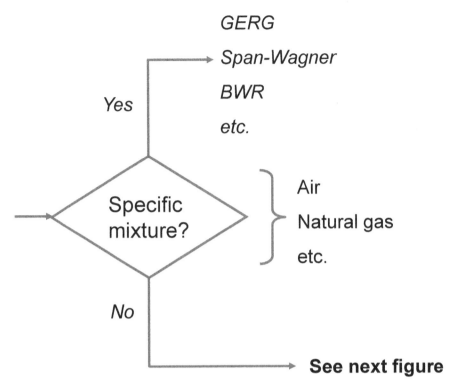

Figure 5.4. *Selection of a thermodynamic model for a mixture (2/4). Case of specific mixtures*

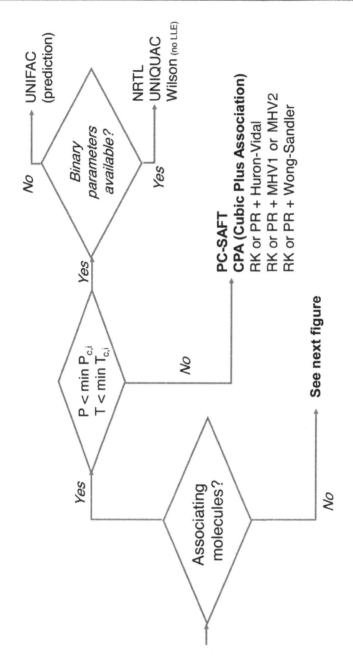

Figure 5.5. *Selection of a thermodynamic model for a mixture (3/4). Case of mixtures that contain associating molecules*

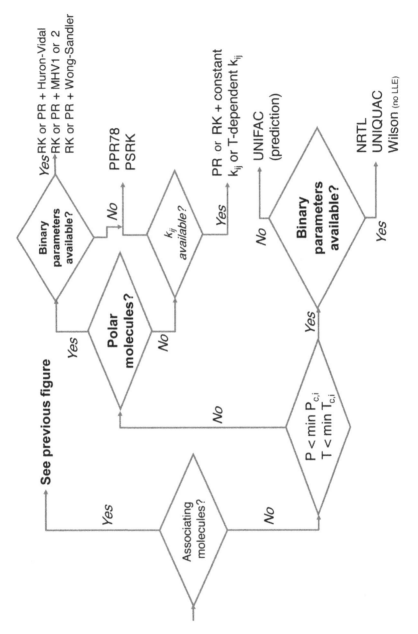

Figure 5.6. *Selection of a thermodynamic model for a mixture (4/4).*
Case of mixtures that do not contain associating molecules

References

Abdoul, W., Rauzy, E., and Péneloux, A. (1991). Group-contribution equation of state for correlating and predicting thermodynamic properties of weakly polar and non-associating mixtures. *Fluid Phase Equilibria*, 68, 47–102. https://doi.org/10.1016/0378-3812(91)85010-R.

Abrams, D.S. and Prausnitz, J.M. (1975). Statistical thermodynamics of liquid mixtures: A new expression for the excess Gibbs energy of partly or completely miscible systems. *AIChE Journal*, 21, 116–128. https://doi.org/10.1002/aic.690210115.

Ahlers, J. and Gmehling, J. (2001). Development of an universal group contribution equation of state. *Fluid Phase Equilibria*, 191, 177–188. https://doi.org/10.1016/S0378-3812(01)00626-4.

Andrews, T. (1869). XVIII. The Bakerian Lecture.—On the continuity of the gaseous and liquid states of matter. *Philosophical Transactions of the Royal Society*, 159, 575–590. https://doi.org/10.1098/rstl.1869.0021.

Avlonitis, G., Mourikas, G., Stamataki, S., and Tassios, D. (1994). A generalized correlation for the interaction coefficients of nitrogen–hydrocarbon binary mixtures. *Fluid Phase Equilibria*, 101, 53–68. https://doi.org/10.1016/0378-3812(94)02554-1.

Benedict, M., Webb, G.B., and Rubin, L.C. (1940). An empirical equation for thermodynamic properties of light hydrocarbons and their mixtures I. Methane, ethane, propane and *n*-butane. *Journal of Chemical Physics*, 8, 334–345. https://doi.org/10.1063/1.1750658.

Boukouvalas, C., Spiliotis, N., Coutsikos, P., Tzouvaras, N., and Tassios, D. (1994). Prediction of vapor–liquid equilibrium with the LCVM model: A linear combination of the Vidal and Michelsen mixing rules coupled with the original UNIFAC and the t-mPR equation of state. *Fluid Phase Equilibria*, 92, 75–106. https://doi.org/10.1016/0378-3812(94)80043-X.

Chen, J., Fischer, K., and Gmehling, J. (2002). Modification of PSRK mixing rules and results for vapor–liquid equilibria, enthalpy of mixing and activity coefficients at infinite dilution. *Fluid Phase Equilibria*, 200, 411–429. https://doi.org/10.1016/S0378-3812(02)00048-1.

Chueh, P.L. and Prausnitz, J.M. (1967). Vapor–liquid equilibria at high pressures: Calculation of partial molar volumes in nonpolar liquid mixtures. *AIChE Journal*, 13, 1099–1107. https://doi.org/10.1002/aic.690130612.

Constantinescu, D. and Gmehling, J. (2016). Further development of modified UNIFAC (Dortmund): Revision and extension 6. *Journal of Chemical and Engineering Data*, 61, 2738–2748. https://doi.org/10.1021/acs.jced.6b00136.

Dahl, S. and Michelsen, M.L. (1990). High-pressure vapor–liquid equilibrium with a UNIFAC-based equation of state. *AIChE Journal*, 36, 1829–1836. https://doi.org/10.1002/aic.690361207.

Fredenslund, A., Jones, R.L., and Prausnitz, J.M. (1975). Group-contribution estimation of activity coefficients in nonideal liquid mixtures. *AIChE Journal*, 21, 1086–1099. https://doi.org/10.1002/aic.690210607.

Gao, G., Daridon, J.-L., Saint-Guirons, H., Xans, P., and Montel, F. (1992). A simple correlation to evaluate binary interaction parameters of the Peng–Robinson equation of state: Binary light hydrocarbon systems. *Fluid Phase Equilibria*, 74, 85–93. https://doi.org/10.1016/0378-3812(92)85054-C.

Graboski, M.S. and Daubert, T.E. (1978). A modified Soave equation of state for phase equilibrium calculations. 2. Systems containing CO_2, H_2S, N_2 and CO. *Industrial and Engineering Chemistry Process Design and Development*, 17, 448–454. https://doi.org/10.1021/i260068a010.

Gross, J. and Sadowski, G. (2001). Perturbed-chain SAFT: An equation of state based on a perturbation theory for chain molecules. *Industrial & Engineering Chemistry Research*, 40, 1244–1260. https://doi.org/10.1021/ie0003887.

Holderbaum, T. and Gmehling, J. (1991). PSRK: A group contribution equation of state based on UNIFAC. *Fluid Phase Equilibria*, 70, 251–265. https://doi.org/10.1016/0378-3812(91)85038-V.

Huron, M.-J. and Vidal, J. (1979). New mixing rules in simple equations of state for representing vapour–liquid equilibria of strongly non-ideal mixtures. *Fluid Phase Equilibria*, 3, 255–271. https://doi.org/10.1016/0378-3812(79)80001-1.

Jaubert, J.-N. and Mutelet, F. (2004). VLE predictions with the Peng–Robinson equation of state and temperature-dependent k_{ij} calculated through a group contribution method. *Fluid Phase Equilibria*, 224, 285–304. https://doi.org/10.1016/j.fluid.2004.06.059.

Jaubert, J.-N. and Privat, R. (2010). Relationship between the binary interaction parameters (k_{ij}) of the Peng–Robinson and those of the Soave–Redlich–Kwong equations of state: Application to the definition of the PR2SRK model. *Fluid Phase Equilibria*, 295, 26–37. https://doi.org/10.1016/j.fluid.2010.03.037.

Jaubert, J.-N., Privat, R., Le Guennec, Y., and Coniglio, L. (2016). Note on the properties altered by application of a Péneloux-type volume translation to an equation of state. *Fluid Phase Equilibria*, 419, 88–95. https://doi.org/10.1016/j.fluid.2016.03.012.

Kato, K., Nagahama, K., and Hirata, M. (1981). Generalized interaction parameters for the Peng–Robinson equation of state: Carbon dioxide *n*-paraffin binary systems. *Fluid Phase Equilibria*, 7, 219–231. https://doi.org/10.1016/0378-3812(81)80009-X.

Kontogeorgis, G.M. and Folas, G.K. (2010). *Thermodynamic Models for Industrial Applications*. John Wiley & Sons, Ltd, Chichester, UK. https://doi.org/10.1002/9780470747537.

Kontogeorgis, G.M., Privat, R., and Jaubert, J.-N. (2019). Taking another look at the van der Waals equation of state–almost 150 years later. *Journal of Chemical & Engineering Data*, 64(11), 4619–4637. https://doi.org/10.1021/acs.jced.9b00264.

Kordas, A., Tsoutsouras, K., Stamataki, S., and Tassios, D. (1994). A generalized correlation for the interaction coefficients of CO_2–hydrocarbon binary mixtures. *Fluid Phase Equilibria*, 93, 141–166. https://doi.org/10.1016/0378-3812(94)87006-3.

Kordas, A., Magoulas, K., Stamataki, S., and Tassios, D. (1995). Methane hydrocarbon interaction parameters correlation for the Peng–Robinson and the t-mPR equation of state. *Fluid Phase Equilibria*, 112, 33–44. https://doi.org/10.1016/0378-3812(95)02787-F.

Kunz, O. and Wagner, W. (2012). The GERG-2008 wide-range equation of state for natural gases and other mixtures: An expansion of GERG-2004. *Journal of Chemical and Engineering Data*, 57, 3032–3091. https://doi.org/10.1021/je300655b.

Le Guennec, Y., Lasala, S., Privat, R., and Jaubert, J.-N. (2016a). A consistency test for α-functions of cubic equations of state. *Fluid Phase Equilibria*, 427, 513–538. https://doi.org/10.1016/j.fluid.2016.07.026.

Le Guennec, Y., Privat, R., and Jaubert, J.-N. (2016b). Development of the translated-consistent tc-PR and tc-RK cubic equations of state for a safe and accurate prediction of volumetric, energetic and saturation properties of pure compounds in the sub- and super-critical domains. *Fluid Phase Equilibria*, 429, 301–312. https://doi.org/10.1016/j.fluid.2016.09.003.

Lohmann, J., Joh, R., and Gmehling, J. (2001). From UNIFAC to modified UNIFAC (Dortmund). *Industrial & Engineering Chemistry Research*, 40, 957–964. https://doi.org/10.1021/ie0005710.

Mathias, P.M. and Copeman, T.W. (1983). Extension of the Peng–Robinson equation of state to complex mixtures: Evaluation of the various forms of the local composition concept. *Fluid Phase Equilibria*, 13, 91–108. https://doi.org/10.1016/0378-3812(83)80084-3.

Maurer, G. and Prausnitz, J.M. (1978). On the derivation and extension of the uniquac equation. *Fluid Phase Equilibria*, 2, 91–99. https://doi.org/10.1016/0378-3812(78)85002-X.

Michelsen, M.L. (1990). A modified Huron–Vidal mixing rule for cubic equations of state. *Fluid Phase Equilibria*, 60, 213–219. https://doi.org/10.1016/0378-3812(90)85053-D.

Michelsen, M.L. and Mollerup, J.M. (2007). *Thermodynamic Models: Fundamentals & Computational Aspects*, 2nd edition. Tie-Line Publications, Holte.

Moysan, J.M., Paradowski, H., and Vidal, J. (1986). Prediction of phase behaviour of gas-containing systems with cubic equations of state. *Chemical Engineering Science*, 41, 2069–2074. https://doi.org/10.1016/0009-2509(86)87123-8.

Nishiumi, H., Arai, T., and Takeuchi, K. (1988). Generalization of the binary interaction parameter of the Peng–Robinson equation of state by component family. *Fluid Phase Equilibria*, 42, 43–62. https://doi.org/10.1016/0378-3812(88)80049-9.

O'Connell, J.P. and Haile, J.M. (2011). *Thermodynamics: Fundamentals for Applications*, 1. Paperback edition (with corr.). Cambridge University Press, Cambridge.

Orbey, H. and Sandler, S.I. (1998). *Modeling Vapor–Liquid Equilibria: Cubic Equations of State and their Mixing Rules*, Cambridge Series in Chemical Engineering. Cambridge University Press, New York.

Péneloux, A., Rauzy, E., and Freze, R. (1982). A consistent correction for Redlich–Kwong–Soave volumes. *Fluid Phase Equilibria*, 8, 7–23. https://doi.org/10.1016/0378-3812(82)80002-2.

Peng, D.-Y. and Robinson, D.B. (1976). A new two-constant equation of state. *Industrial and Engineering Chemistry: Fundamentals*, 15, 59–64. https://doi.org/10.1021/i160057a011.

Pina-Martinez, A., Privat, R., Jaubert, J.-N., and Peng, D.-Y. (2019). Updated versions of the generalized Soave α-function suitable for the Redlich–Kwong and Peng–Robinson equations of state. *Fluid Phase Equilibria*, 485, 264–269. https://doi.org/10.1016/j.fluid.2018.12.007.

Prausnitz, J.M. (1985). Equations of state from van der Waals theory: The legacy of Otto Redlich. *Fluid Phase Equilibria*, 24, 63–76. https://doi.org/10.1016/0378-3812(85)87037-0.

Prausnitz, J.M., Lichtenthaler, R.N., and de Azevedo, E.G. (1999). *Molecular Thermodynamics of Fluid-Phase Equilibria*, 3rd edition. Prentice-Hall International Series in the Physical and Chemical Engineering Sciences. Prentice Hall PTR, Upper Saddle River, NJ.

Privat, R. and Jaubert, J.-N. (2012). Discussion around the paradigm of ideal mixtures with emphasis on the definition of the property changes on mixing. *Chemical Engineering Research and Design*, 82, 319–333. https://doi.org/10.1016/j.ces.2012.07.030.

Privat, R. and Jaubert, J.-N. (2013). Classification of global fluid-phase equilibrium behaviors in binary systems. *Chemical Engineering Research and Design*, 91, 1807–1839. https://doi.org/10.1016/j.cherd.2013.06.026.

Privat, R., Jaubert, J.-N., and Le Guennec, Y. (2016). Incorporation of a volume translation in an equation of state for fluid mixtures: Which combining rule? Which effect on properties of mixing? *Fluid Phase Equilibria*, 427, 414–420. https://doi.org/10.1016/j.fluid.2016.07.035.

Qian, J.-W., Privat, R., Jaubert, J.-N., and Duchet-Suchaux, P. (2013). Enthalpy and heat capacity changes on mixing: Fundamental aspects and prediction by means of the PPR78 cubic equation of state. *Energy Fuels*, 27, 7150–7178. https://doi.org/10.1021/ef401605c.

Redlich, O. and Kwong, J.N.S. (1949). On the thermodynamics of solutions. V. An equation of state. Fugacities of gaseous solutions. *Chemical Reviews*, 44, 233–244.

Renon, H. and Prausnitz, J.M. (1968). Local compositions in thermodynamic excess functions for liquid mixtures. *AIChE Journal*, 14, 135–144. https://doi.org/10.1002/aic.690140124.

Rowlinson, J.S. (1971). *Liquids and Liquid Mixtures*, 2nd edition. Butterworth, London.

Sandler, S.I. (ed.) (1994). *Models for Thermodynamic and Phase Equilibria Calculations, Chemical Industries*. Dekker, New York.

Sandler, S.I. (2017). *Chemical, Biochemical and Engineering Thermodynamics*, 5th edition. Wiley, Hoboken, NJ.

Smith, J.M., Van Ness, H.C., Abbott, M.M., and Swihart, M.T. (2018). *Introduction to Chemical Engineering Thermodynamics*, 8th edition. McGraw-Hill Education, New York, NY.

Soave, G. (1972). Equilibrium constants from a modified Redlich–Kwong equation of state. *Chemical Engineering Science*, 27, 1197–1203. https://doi.org/10.1016/0009-2509(72)80096-4.

Span, R. and Wagner, W. (2003a). Equations of state for technical applications. I. Simultaneously optimized functional forms for nonpolar and polar fluids. *International Journal of Thermophysics*, 24, 1–39. https://doi.org/10.1023/A:1022390430888.

Span, R. and Wagner, W. (2003b). Equations of state for technical applications. II. Results for nonpolar fluids. *International Journal of Thermophysics*, 24, 41–109. https://doi.org/10.1023/A:1022310214958.

Span, R. and Wagner, W. (2003c). Equations of state for technical applications. III. Results for polar fluids. *International Journal of Thermophysics*, 24, 111–162. https://doi.org/10.1023/A:1022362231796.

Stryjek, R. (1990). Correlation and prediction of VLE data for *n*-alkane mixtures. *Fluid Phase Equilibria*, 56, 141–152. https://doi.org/10.1016/0378-3812(90)85099-V.

Tassios, D.P. (1993). *Applied Chemical Engineering Thermodynamics*. Springer, Berlin, Heidelberg. https://doi.org/10.1007/978-3-662-01645-9.

Twu, C.H., Bluck, D., Cunningham, J.R., and Coon, J.E. (1991). A cubic equation of state with a new alpha function and a new mixing rule. *Fluid Phase Equilibria*, 69, 33–50. https://doi.org/10.1016/0378-3812(91)90024-2.

Twu, C.H., Coon, J.E., and Cunningham, J.R. (1995a). A new generalized alpha function for a cubic equation of state. Part 1: Peng–Robinson equation. *Fluid Phase Equilibria*, 105, 49–59. https://doi.org/10.1016/0378-3812(94)02601-V.

Twu, C.H., Coon, J.E., and Cunningham, J.R. (1995b). A new generalized alpha function for a cubic equation of state. Part 2: Redlich–Kwong equation. *Fluid Phase Equilibria*, 105, 61–69. https://doi.org/10.1016/0378-3812(94)02602-W.

Valderrama, J.O., Ibrahim, A.A., and Cisternas, L.A. (1990). Temperature-dependent interaction parameters in cubic equations of state for nitrogen-containing mixtures. *Fluid Phase Equilibria*, 59, 195–205. https://doi.org/10.1016/0378-3812(90)85034-8.

Valderrama, J.O., Arce, P.F., and Ibrahim, A.A. (1999). Vapour–liquid equilibrium of H_2S–hydrocarbon mixtures using a generalized cubic equation of state. *Canadian Journal of Chemical Engineering*, 77, 1239–1243. https://doi.org/10.1002/cjce.5450770622.

Van der Waals, J.D. (1873). *On the Continuity of the Gaseous and Liquid States*. Leiden.

Vidal, J. (2003). *Thermodynamics: Applications in Chemical Engineering and the Petroleum Industry*. Editions Technip, Paris.

Voutsas, E., Magoulas, K., and Tassios, D. (2004). Universal mixing rule for cubic equations of state applicable to symmetric and asymmetric systems: Results with the Peng–Robinson equation of state. *Industrial & Engineering Chemistry Research*, 43, 6238–6246. https://doi.org/10.1021/ie049580p.

Voutsas, E., Louli, V., Boukouvalas, C., Magoulas, K., and Tassios, D. (2006). Thermodynamic Property Calculations with the Universal Mixing Rule for EoS/GE Models: Results with the Peng–Robinson EoS and a UNIFAC Model. *Fluid Phase Equilibria*, 241, 216–228. https://doi.org/10.1016/j.fluid.2005.12.028.

Wilson, G.M. (1964). Vapor–liquid equilibrium. XI. A new expression for the excess free energy of mixing. *Journal of the American Chemical Society*, 86, 127–130. https://doi.org/10.1021/ja01056a002.

Wong, D.S.H. and Sandler, S.I. (1992). A theoretically correct mixing rule for cubic equations of state. *AIChE Journal*, 38, 671–680. https://doi.org/10.1002/aic.690380505.

Xu, X., Jaubert, J.-N., Privat, R., and Arpentinier, P. (2017). Prediction of thermodynamic properties of alkyne-containing mixtures with the E-PPR78 model. *Industrial & Engineering Chemistry Research*, 56, 8143–8157. https://doi.org/10.1021/acs.iecr.7b01586.

Index

Printed in the United States
by Baker & Taylor Publisher Services